EVERY AMERICAN AN INNOVATOR

EVERY AMERICAN AN INNOVATOR

HOW INNOVATION BECAME A WAY OF LIFE

MATTHEW WISNIOSKI

The MIT Press
Cambridge, Massachusetts
London, England

in association with
The Lemelson Center
Smithsonian Institution
Washington, D.C.

The MIT Press
Massachusetts Institute of Technology
77 Massachusetts Avenue, Cambridge, MA 02139
mitpress.mit.edu

Publication of this open monograph was the result of Virginia Tech's participation in TOME (Toward an Open Monograph Ecosystem), a collaboration of the Association of American Universities, the Association of University Presses, and the Association of Research Libraries. TOME aims to expand the reach of long-form humanities and social science scholarship including digital scholarship. Additionally, the program looks to ensure the sustainability of university press monograph publishing by supporting the highest quality scholarship and promoting a new ecology of scholarly publishing in which authors' institutions bear the publication costs.

Funding from Virginia Tech made it possible to open this publication to the world. www .openmonographs.org

The MIT Press would like to thank the anonymous peer reviewers who provided comments on drafts of this book. The generous work of academic experts is essential for establishing the authority and quality of our publications. We acknowledge with gratitude the contributions of these otherwise uncredited readers.

This book was set in Stone Serif and Stone Sans by Westchester Publishing Services. Printed and bound in the United States of America.

Library of Congress Cataloging-in-Publication Data

Names: Wisnioski, Matthew H., 1978- author.
Title: Every American an innovator : how innovation became a way of life / Matthew Wisnioski.
Other titles: How innovation became a way of life
Description: Cambridge, Massachusetts : The MIT Press, [2025] | Series: Lemelson Center Studies in invention and innovation series | Includes bibliographical references and index.
Identifiers: LCCN 2024031647 (print) | LCCN 2024031648 (ebook) | ISBN 9780262550734 (paperback) | ISBN 9780262381062 (epub) | ISBN 9780262381079 (pdf)
Subjects: LCSH: Technological innovations—United States—History. | Technological innovations—Social aspects—United States—History. | United States—Social life and customs.
Classification: LCC T173.4 .W57 2025 (print) | LCC T173.4 (ebook) | DDC 609.73—dc23 /eng/20241206
LC record available at https://lccn.loc.gov/2024031647
LC ebook record available at https://lccn.loc.gov/2024031648

10 9 8 7 6 5 4 3 2 1

EU product safety and compliance information contact is: mitp-eu-gpsr@mit.edu

Everyone talking together makes something that seems like me but is not me. Everyone doing things in the world makes me.

—Kim Stanley Robinson, *Ministry for the Future*

CONTENTS

SERIES FOREWORD

Think. Think different. Imagination at work. Advertising slogans from IBM, Apple, GE, and many other companies have trumpeted a commitment to innovation. Indeed, invention and innovation have long been recognized as transformational forces in American history, not only in technological realms but also in politics, society, and culture. In recent decades, how did this push to innovate become not just accepted but expected?

In *Every American an Innovator*, author Matthew Wisnioski explores how the imperative to innovate—a concept that he points out "barely registered as a cultural touchstone" before the 1950s—became "ingrained in our institutions, our educational system, and our beliefs about ourselves." Along the way, innovation gained advocates, as well as critics, and the ranks of innovators have become inclusive enough to embrace not only Steve Jobs but also Girl Scouts selling cookies emblazoned with "I am an innovator."

Since 1995, the Smithsonian's Lemelson Center has been investigating the history of invention and innovation from an interdisciplinary perspective. Books in the Lemelson Center Studies in Invention and Innovation extend this work to enhance public understanding of humanity's inventive impulse. Authors in the series raise new questions about the work of diverse inventors and the technologies they create, while stimulating cross-disciplinary dialogue. By opening channels of communication between the various disciplines and sectors of society concerned with technological innovation, the Lemelson Center Studies advance scholarship in the history of technology, engineering, science, architecture, the arts, and related fields and disseminate it to a general interest audience.

Joyce Bedi, Arthur Daemmrich, and Eric S. Hintz
Series editors, Lemelson Center Studies in Invention and Innovation

PREFACE

Where do ideas come from and how do they alter the world as they travel?

The seed of this book began with a chance discovery two decades ago. As a graduate student, I would roam the basement of Princeton's Firestone Library where ideas go to be forgotten. One day, in the oversized section, which often held the best treasures, I encountered an obscure magazine from the 1960s that described our present world with remarkable accuracy; so much so that I didn't know how to place it. The magazine, *Innovation*, recounted experiments in making technology serve humanity. It lionized the mavericks who risked their careers to do so, chronicling their rewards and failures, and it analyzed with passion the contentious issues of knowledge, equity, and democracy at stake. *Innovation*'s authors fashioned themselves as new kinds of experts during a moment of national turmoil to assert that world events demanded transformations in our institutions and our inner selves.

I have spent more than a decade studying the growth and evolution of the innovation culture imagined by those experts. Generous support from the National Science Foundation and the Lemelson Center for the Study of Invention and Innovation, two institutions that figure prominently in this book, made my search possible. *Every American an Innovator* shares the revelations, stories, and insights that I've found in archives across the country; thousands of policy reports, journals, and popular ephemera; as well as interviews and observations with current practitioners.

When I began, innovation had nearly unquestioned status as a social good. I wanted to know why that was the case. This universal support was at odds with the contentious role that the related idea of *technology* has played in American history. In past work, I explored how the societal rifts of the 1960s shaped competing beliefs about technology's value and consequences. Innovation in contrast emerged as a shared value in an age of fracture.

As luck would have it, I dug into the archives just as scholars and activists started to challenge innovation's supremacy. These critics, many of whom are friends and collaborators, raised important concerns about innovation culture that I share. But this "Techlash," as journalists came to call it, only increased my curiosity about how and why we value or reject innovation as a societal ideal.

In search of answers, I joined a series of innovation initiatives as a critical participant. I contributed to a university center that merges art and technology. I collaborated on a multiyear reform project to "revolutionize" engineering education. I then helped lead a visioning exercise that laid the groundwork for a billion-dollar Innovation Campus that I later publicly questioned. With colleagues at the Smithsonian, I staged a dialogue among prominent champions, critics, and reformers of innovation culture.[1] As a result, I have personal relationships and vested interests with many people and organizations that I analyze in these pages. This may not be the best strategy for keeping friends and patrons. Nonetheless, I have tried to tell their stories in a way that they would recognize as fair and that illuminates the motivations, implications, and outcomes of their varied projects.

This book is the history of a society altered by its pursuit of innovation as an organizing theme and an inquiry into its recent splintering. I write as a historian motivated to understand how ideas travel and become dominant, as well as a conflicted reformer who has sought to bend their path. I trace how a concept that prior to World War II barely registered as a cultural touchstone became ingrained in our institutions, our educational system, and our beliefs about ourselves. It is the story of how, for better and for worse, innovation became a way of life.

Blacksburg, Virginia
September 2024

1 MAKE THINGS BETTER®

Innovation is certainly a "buzz-word" today. Everyone likes the idea; everyone is trying to "innovate"; and everyone wants to do better at it tomorrow.
—Jack A. Morton, 1971

On the fourteenth floor of a downtown high-rise, Pittsburgh's change agents ideate renewal. A multicultural group of eighteen teachers and principals work to transform the city's struggling public schools. Their goal: create the next generation of innovators.[1]

The mood in the whiteboard-clad room is hopeful and collaborative. The session is the first in a multiyear series addressing everything from creative technologies to social justice. "We're ready to make the leap," one teacher declares. "I'm not going to be alone," another adds. Their blueprint is decentralized and personal. "Top down" approaches are "foolish" says Jeanne Perlman of the Pittsburgh Foundation, the initiative's sponsor, "because we know as human beings that children are different."

The meeting is decidedly future-oriented. What is "not now," its instructor Bill Lucas assures, "will soon be." Lucas is the cofounder of the LUMA Institute. LUMA, which stands for "Looking, Understanding, and Making," is in the business of innovation expertise. Lucas's purpose is to equip teachers and students alike with the "innovative tools" to become "the very best that they can be." His consulting firm has surveyed and distilled over nine hundred techniques from a half-century of the world's innovators. It

FIGURE 1.1
LUMA's Change Agents Workshop. Photo by Joshua Franzos. Courtesy of Pittsburgh Foundation.

has applied these techniques for tech giants such as Google and Autodesk but also McDonald's, the American Cancer Society, the Harvard Business School, and even the White House. Those who cannot afford LUMA's workshop fees, or who live afield of the global hubs where it offers courses, can purchase a deck of Innovating for People planning cards, a pocket-sized toolkit for "bringing new and lasting value to the world."[2]

To innovate is the twenty-first century's defining imperative. Nation states develop *innovation ecosystems* to compete in a fast-moving global economy. Cities build *innovation districts* to attract venture capital and start-ups. Companies compete for market share and talent by claiming that their cultures are the most innovative. Universities aim to turn scientific discoveries into breakthrough technologies. Foundations invest in social entrepreneurs to confront systemic inequity. School systems seek to shake-up entrenched bureaucracies to mold young minds.

Across the United States and around the world, the imperative to innovate dominates understandings of the past, diagnoses of the present, and visions of the future. Advocates for this capacious imperative share a common belief, which LUMA has trademarked—to innovate is to Make Things Better®.

TRANSFORMATIONAL PROMISES

This book investigates where the United States' demand for innovation originates. It asks: What has led engineers, schoolteachers, and presidents alike to declare innovation the solution? How has the meaning of innovation evolved as an expanding range of Americans embraced it? What kind of *culture*—manifest in organizations, practices, and patterns of living—have innovation's advocates created; and what have been the consequences of this transformation?

The answers to these historical questions are vital for a society that has started to reconsider its relationship with innovation. For decades, corporate executives and government leaders have described innovation as a driver of economic and social progress, insisting that everyone must innovate for a better future. However, critics increasingly challenge the clamor for innovation. Journalists document how start-up companies have become the monopolistic corporations they once disrupted. Television and film, which previously valorized Silicon Valley, now satirize greed-driven innovators. Scholars in my own field of science and technology studies confront innovation boosters with the fervor of a religious reformation. The philosopher Langdon Winner, for example, has described innovation culture as a cult, while my colleague Lee Vinsel characterized LUMA's techniques as akin to syphilis that "rots your brains."[3]

What could possibly be wrong with coming up with new ideas? To answer this question, we must look at how the demand for innovation came about and what its consequences are. Innovation, in its ubiquity, may seem like just a buzzword or a natural expression of the spirit of our time. However, today's visions of innovation are the result of a relentless effort by people in traceable networks to promote certain ideas and of the convergence between those ideas and changing societal goals. Their collective work has resulted in new institutions, new technologies, and new ways of living.

The history of this societywide demand for innovation is both older and more recent than commonly understood. Advocates often view innovation as an entirely new phenomenon whose secrets they discovered, or as a phenomenon that has always existed. From this perspective, history illuminates the success of specific technologies and the innovators who created them. In contrast, academic historians demonstrate that the concept of "innovation" dates to ancient Greece and are quick to point out that America's current infatuation is not unique.[4]

The innovation imperative is also culturally and geographically specific. To be sure, innovation is not just an American obsession. From the European Union to China, India, Colombia, and Nigeria, advocates work to build innovation cultures with similar goals and blueprints.[5] Still, as this book will show, today's innovation culture had specific origins in the United States in the 1950s and 1960s. Moreover, in the ensuing decades, the United States played an outsized role in spreading the specific models and broad visions of innovation that have been appropriated throughout the world.

To understand innovation's cultural power and why the United States is central to its rise, I follow the concept's meaning in use. Innovation is a protean idea that links global economic change to the inner life of school children. The concept does not have a universal or fixed meaning. Nor was there a specific moment in the past when innovation culture sprung forth wholly formed. The idea of innovation did not flow in linear fashion from the minds of economists or the research parks of Silicon Valley. Rather it arose in multiple locations and it has been taken up and refashioned in surprising ways over seventy-five years.

Following the evolution of innovation's meaning highlights persistent tensions in American culture from World War II to today. Innovation is at the root of a series of transformational promises across an era of scientific and technological change, cultural and economic globalization, and political conflict. Americans have called upon innovation as a path forward to better futures that overcome political divisions and broaden participation in a technological society. Those promises, however, have been intertwined with competing ideologies and fears of national decline.

Before investigating where innovation's promises and tensions originated, it helps to understand how they operate today.

HUMANITY AND TECHNOLOGY

Pittsburgh—which appears throughout this book as a crucible of innovation culture—is an appropriate place to begin. There are signs everywhere of a city whose fortunes have been transformed by technological innovation.

FIGURE 1.2

Pittsburgh TechMap, 2017–2018. Courtesy of Pittsburgh Technology Council.

In 2006, Google built an office on the grounds of a former industrial bakery not far from the East Liberty neighborhood where farm-to-table restaurants sprouted from abandoned buildings. In 2015, the on-demand taxi company Uber arrived, poaching nearly the entirety of Carnegie Mellon's (CMU) automated vehicle group, only months after forming a research partnership with the world-renowned university.[6] A few years later, a systematic survey by the city's Technology Council, titled "Welcome to Nerdvana," identified nearly five hundred companies in the region's innovation ecosystem.[7] Now, with over twenty artificial intelligence start-ups clustered around Bakery Square, boosters welcome investors to stroll down "AI Avenue."[8]

Seeking Nerdvana is an inquiry into the right relationships between humanity and technology. For many people, *innovation* is synonymous with *technology*; or more accurately, rapid technological change. Innovation is the computer revolution, five billion mobile phone users, and the sequencing of the human genome. It is the promise and fear of self-driving cars and artificial intelligence. In the most deterministic interpretations, technology drives history. It has its own logic to which we are captive.[9] The primary role that humans play with respect to technology is as discoverers and inventors. Innovation is distinct from technology and evokes a more complex relationship with humanity.

Economic growth is at the center of innovation's promises. Leading economists assert that technological change, rather than labor or capital, is the key source of societal wealth. This premise reinforces an expansive understanding of society and politics. Innovation, and therefore wealth, springs from the agglomeration of research universities, start-up companies, and regional coalitions.[10] Since the 1970s, policymakers have stressed the need to transfer and translate the work of scientists into the real world. Economic geographers likewise provide explanations of what made the Bay Area, Boston's Route 128, and North Carolina's Research Triangle Park so successful.[11]

Still, innovation clearly means more than economic growth through technology. Workshops like those held at LUMA encourage teachers and students to identify authentic problems and to achieve social justice in their communities. These experts coach that ordinary people should not accept the status quo and that they can work to change it. Advocates use techniques that blend technological and social processes to challenge structures of power and authority and to seek creative solutions. Innovation's champions assert that these techniques and mindsets can change communities, nations, and even humanity at large. Innovation, in short, is a tool for empowering humanity.

NEW MEN AND EVERY GIRL

Just as innovation describes new relationships between technology and humanity, it creates new identities for people to aspire. The persona of the *innovator* occupies a special place in the American imagination. Innovators are the "change makers" who approach entrenched problems in new ways, harnessing their creativity to make technology serve human needs.

This view of innovators draws on similar tropes ascribed to inventors in the nineteenth century and even to Renaissance artists; however, the qualities of innovators are novel to the twentieth and twenty-first centuries. In the 1930s, the economist Joseph Schumpeter captured an important element of what defines innovators, valorizing the entrepreneurial risk-taking and persistence of capitalism's "New Men." Schumpeter helped establish the contours of the innovator's persona, although it is largely a post–World War II invention. In the early 1960s, media portrayals identified innovators as new kinds of scientists and technologists who differed from the genius physicists behind the atom bomb or the company men who filled the ranks of large corporations. They were a select group who transgressed boundaries between science, technology, and society. Innovators were iconoclasts who saw the future *and* who had the skills to make it real.

In recent decades, innovation's patron saint has been Steve Jobs, whose place in the iconography of innovation began as soon as Apple started selling computers. In his career, Jobs gave hundreds of interviews and speeches, and his mythic stature has only grown since his death in 2011. In Walter Isaacson's *Steve Jobs*, among the bestselling biographies of all time, the tech innovator's stature matches that of Benjamin Franklin at the republic's founding or Thomas Edison during the age of American invention.[12] A complex, even tragic figure, the innovator is passionate and cold, an artist and an engineer, a risk-taker and a calculating planner, driven at a young age but playful and curious throughout life. Above all, the innovator is a visionary who turns dreams into reality.

But who can be an innovator? Today, the answer seems to be *everyone*. As LUMA's workshops attest, there are myriad efforts to train a diverse nation of innovators. From camps aimed at elementary schoolers to the National Science Foundation's Innovation Corps, innovation is an activity in which Americans of every race, class, and gender are encouraged to participate. Recently, the Girl Scouts introduced a new line of cookies celebrating creativity, grit, and entrepreneurship. "I am an innovator," the lemon biscuit declares.

Efforts like the Girl Scouts' campaign to expand innovation from turtleneck- and hoodie-wearing white men to children of diverse, intersectional backgrounds have historical precedent. Critics of innovation may be surprised to learn that feminist calls for empowerment and movements for racial equity in science and technology have shaped innovation culture since the early 1980s. These values, in turn, resonate with popular theories of democratic innovation and have developed in tandem with demands for a national innovation workforce.

Whether innovators are an avant-garde or whether everyone can participate in innovation is among the most vexing questions of innovation culture. It is a practical issue over where educational resources ought to be spent, as well as a conceptual one over whose contributions have the most value to society. Asking who gets to innovate also brings to the forefront whether innovation is a collective enterprise aimed at making things better for *everyone*, or whether the valorization of innovators leads to a situation in which *every one* competes individually with each other in a global market.

EXPERTISE AND IMAGINATION

Innovators are the most visible symbols of innovation culture, but a second and overlapping group of change agents is just as important in fueling the imperative to innovate. These are the *innovation experts* whose theories, techniques, and mindsets are at the core of innovation culture. LUMA is one example of a global industry that offers insights into what innovation is, how to make it, and how to capitalize on that knowledge.

A few of these innovation experts have become famous. Harvard Business School professor Clayton Christensen, for example, gained notoriety in the early twenty-first century for his theory of disruptive innovation. Richard Florida likewise commanded hefty consulting fees for asserting that innovation spawned a new social class that determines which regions succeed and which fail.[13]

Most innovation experts, however, work outside of the limelight. They remain anonymous in part because the label "innovation expert" has only been in use since the 1990s. But the cultivation of expertise for understanding and navigating disruptive change is far older and its reach is profound. Starting in the 1950s, an eclectic group of economists, philosophers, sociologists, journalists, bureaucrats, business consultants, and teachers claimed authority over the processes of innovation. These experts developed methods and practices for enhancing innovative activity. They also created powerful images of innovation's role in historical change and the innovators that would bring it about. They created new organizations and worked to reform the nation's businesses, government agencies, universities, and elementary schools.

Pittsburgh's LUMA reveals the core patterns of innovation expertise. Its parent company, MAYA, was founded in 1988 by three CMU dropouts.[14] Unlike the stereotypical start-up, its founders weren't precocious teenagers. They included two former department heads with a combined forty years of work as research scientists. Together, they combined the expertise of an industrial designer, a computer scientist, and a cognitive psychologist. Along with a small number of peers, including the more famous IDEO design company, they pioneered the field of human-centered design. MAYA's first client was the now defunct Digital Equipment Corporation, considered the beacon of American innovation during the computer revolution of the 1960s. Contracts with the Department of Defense followed. MAYA then expanded from work at the human–computer interface, to product design, organizational culture, and creative thinking. With LUMA, it discovered that innovation expertise itself was a viable product. But LUMA's roots go deeper. In the 1950s and 1960s, CMU rose to national prominence through the merger of a technical institute and a liberal arts college; its president would serve as a national science advisor and the university would host one of the country's first innovator incubators.

My field of science and technology studies has also been a major contributor of innovation expertise. In the 1960s, concerned scientists, engineers, humanists, social scientists, and citizens came together to challenge the dominance of the military and corporations in technological development. They asked how the making of science and technology could be more democratic and how its uses could better serve society. Science and technology studies scholars have explained what innovation is, where it comes from, and how to make better innovators. Their prescriptions often echo LUMA's optimism as they seek to make innovation *inclusive, responsible*, and *just*.

Innovation expertise has proven to be a robust and widely shared invention, traveling from government agencies and executive boardrooms to nearly every sector of American society. But is it effective? Well-meaning reformers attempt to disrupt traditional bureaucracies and failing systems with optimistic mindsets and technological solutions. Instead, they often reproduce the status quo. Initiatives to produce innovators encourage "creative confidence" in language that resembles the self-help movement of an earlier industrial era.[15] Finally, some critics see innovation expertise as a confidence game in which swindlers sell a pernicious form of "innovation speak," inflicting individual and collective damage on how we live today.[16]

ASPIRATION AND ANXIETY

Pittsburgh again is a useful place to begin to understand the tension between innovation culture's promises and its realities. Myths of innovation are tales of triumphant outliers. Thomas Edison and his Menlo Park wizards. The Bell Laboratories "idea factory" where Nobel Prize winners invented the transistor and discovered the scientific foundations of the information age. MIT as the world's preeminent educator of innovators and entrepreneurs. The exponential growth of Silicon Valley. The rise of Apple, Amazon, Facebook, and Google. These tales of heroic innovation are exhilarating, powerful, and occasionally true. But a focus only on winners misses an important feature about how innovation culture operates—perhaps the most important feature of all.

In contrast to San Francisco and Boston, Pittsburgh is both a textbook example of American decline and a hopeful tale of national renewal. Once the world's largest producer of steel, a combination of deregulation, automation, white flight, and global competition in the 1970s and 1980s resulted in over 150,000 lost manufacturing jobs and a reduction of 300,000 residents.[17] Consequently, for almost half a century, the city has been the site of one reform project after another. In 1999, for example, a diverse collection of politicians, business leaders, nonprofits, and entrepreneurs coordinated with a Palo Alto consultancy firm to identify the "secrets of innovative regions."[18] As these investments paid dividends, promoters tout Pittsburgh as one of America's "most livable" cities with the creative spirit and entrepreneurial vitality of Silicon Valley, but at an affordable and balanced scale.[19]

Pittsburgh showcases innovation culture's history of aspiration. Boosters promise that through innovation we can change our course for the better. That old ways can be disrupted. That progress is not just desirable, but attainable. These aspirations drive the quest for innovation around the world. Policymakers seek to recreate the history of successful cases to become the next Jobs, the next MIT, the next Silicon Valley.

But hope's dark mirror is anxiety. The demand for innovation in the United States comes in response to fears of economic, societal, and even moral decline. Over the past half-century, as corporations offshored manufacturing and hollowed out the nation's Rust Belt cities, the demand for innovation only increased. Calls to innovate have ranged from the ambivalent command to compete in a "flat world" to the menacing danger of a "gathering storm."[20]

Scholars describe this anxious worldview as an *innovation deficit model*. Calls for innovation demand that we think creatively and improve our productivity lest we fall behind competitors and history itself. This fear leads to imitation rather than innovation. Pittsburgh's Nerdvana map, for instance, emulates those drawn in Silicon Valley over fifty years ago. Nearly every American city today has a similar council and a similar map. While Pittsburgh claims its share of successes, others, such as Louisiana's erstwhile attempt to build a Silicon Bayou in the 1980s, failed to catch fire.[21]

The tension between aspiration and anxiety is a defining feature of innovation culture. Helga Nowotny, a philosopher who helped author the European Union's innovation strategy, described this condition as innovation's "fragile future." Innovation, she writes, is based on a "fundamental societal consensus that is nevertheless brittle and must be constantly renegotiated."[22]

The brittleness of innovation as a concept and as a culture is increasingly apparent. After decades of chasing aspirations, critics now question whether the transformations promised in innovation's name have been met. Glowing portraits of innovators have also lost their shine. Current profiles are more likely to vilify innovators than they are to cast them as gods. Mark Zuckerberg and Elon Musk have become antiheroes.

Critics of innovation culture find a unitary explanation for a litany of its specific shortcomings in the rise of neoliberalism. As a political philosophy, neoliberalism is the premise that economic markets drive society.[23] In the neoliberal worldview, science and technology are financial commodities and everyone is their own entrepreneur. The ultimate consequence is the erasure of society as a collective project. Neoliberalism, its critics assail, is capitalism's triumph over human life brought about through a dangerous interpretation of the relationship between technology and humanity—that of innovation.

THE INNOVATION OF INNOVATION

How can one concept contain so much? The question cannot be answered in purely linguistic terms nor in abstractions of political philosophy. Certainly, answers will not be found in the deterministic march of technology. Neither can the competing and contradictory tensions fueling innovation culture and its discontents be attributed only to entrepreneur heroes or neoliberal villains. Similarly, we cannot praise or lay the blame on any one political faction. Belief in the transformational power of innovation has transcended those partisan divisions and made strange bedfellows among social progressives and libertarian conservatives throughout innovation culture's ascendancy across an era of widening partisanship. It has stemmed not only from Silicon Valley tech giants and Cambridge biomedical research laboratories; Midwestern extension agents long ago worked out central themes of innovation culture.

To accurately account for innovation culture's rise to dominance and its recent faltering, I explore how the meanings, practices politics, and myths of innovation developed in concert. That process of assemblage has come to encompass thousands of organizations and millions of people. But its key protagonists were a small group of innovation experts. Innovation experts played a profound role in binding together competing beliefs about innovation and society including the boundaries between engineering and science, the sources of social change, the proper relationship between government and private enterprise, the origins of human creativity, and who gets to participate in designing the future. Throughout the book, I profile the best-known of these gurus but also the anonymous experts who worked behind the scenes.

One of the most important inventions of the innovation experts was the identification of the *innovator* as a desirable social type. The personification of innovation as a change agent dedicated to making things better gave a human face to historical forces. At the same time, it directed those forces inward, promising that through new practices and mindsets, it was possible to achieve one's best self via highly specific visions of the good.[24]

This persona of the innovator is the organizing object of the book. I trace the spread and evolution of innovator ideals from elite professionals to children over more than half a century. My focus goes beyond the analysis of cultural and intellectual representations. The innovator was what scholars refer to as an "imaginary," but it was one that innovation experts set out to make real. In the chapters that follow, I show how as innovation experts worked to cultivate a nation of innovators, they reshaped the structures and practices of science, technology, and society in the United States.

In part I, "Seedlings and Incubators," I document how different expert communities in the 1950s transformed an ancient idea into a modern theory of change with the innovator as its leading agent. I then explore the fledgling efforts of innovation experts in the 1960s and 1970s to implement their visions inside government, industry, and universities. At a historical moment marked by diminishing faith in Cold War democracy and seemingly uncontrollable technological change, these experts were driven by an almost utopian belief in innovation's potential to improve humanity. Their attempts to champion innovation included a federal award on par with the Nobel Prize, the establishment of a national network of switched-on entrepreneurs who claimed to combine capitalism with countercultural free-thinking and social justice, and programs to train new scientists and engineers to serve the domestic economy instead of the military–industrial complex.

In part II, "Deficits and Dreams," I document how innovation became the nationalist and global imperative that we experience today. Focusing on the late 1970s to the early 2000s, I first trace the rise of a muscular innovation policy. I next show how an ideology spawned by fears about industrial productivity ironically evolved into a quest to build tolerant communities. I then explore how feminist and antiracist experts worked to expand who counted as innovators by seeking to "empower" women and minorities to remake innovation culture. Finally, I document how philanthropic foundations partnered with governments, universities, companies, and other nonprofits to make all the nation's children into innovators.

In part III, "Self-Reckoning," I close by surveying the height of innovation exuberance during the administration of President Barack Obama and the anti-innovation backlash of recent years. I document how tensions running throughout the decades-long expansion of innovation culture contributed to a series of progressive reform movements, including the responsible innovation, design justice, and maintainers movements. I also investigate the rise of conservative interpretations of innovation culture that celebrated the value of contrarian and transgressive ideas.

My purpose is neither to valorize innovation as a distinctly American achievement nor to debunk the pursuit of innovation as a delusion. Rather, by exploring the contradictory impulses that have fueled innovation culture, we can see with greater clarity why innovation came to be a universal good, who we envision future generations to be, how we should transform society to make them, why individuals and organizations struggle with the limitations that vision creates, why they now are a source of discord, and how we might imagine more equitable foundations for social change. Whether we hope to harness innovation's social power or to challenge it, we need to understand how innovation became a way of life in the first place.

I SEEDLINGS AND INCUBATORS

2 WHEN INNOVATION WAS NEW

How many of us can really recognize in the vast clutter of modern life the seed-lings of new ideas and new ways that will shape the future?
—John Gardner, 1964

On a cornfield in Collins, Iowa, in 1955, amid breakthroughs in digital computing, atomic physics, and molecular genetics, a young farmer's son established the fundamental principles of innovation. Everett Rogers, a PhD candidate at Iowa State, found that farmers did not inevitably use a new technology, nor did they base their decisions about that technology simply on economic rationality. He figured this out by asking nearly 150 farmers about the operational details of their farms, which neighbors the farmers respected most, what newspapers they read, and whether their wives were related to their landlords.[1]

Rogers's proximate interest was the farmers' use—or nonuse—of 2,4-D weed spray. In low doses 2,4-Dichlorophenoxyacetic acid, which had been developed as a chemical warfare agent during World War II, stunted the growth of broadleaf plants but did not affect narrowleaf crops such as wheat, corn, and rice. In other words, the organic molecule was an ideal weed killer. First marketed in 1945 as *Weedone* by the American Chemical Paint Company, 2,4-D had grown to be the most utilized herbicide in the world.[2]

Rural sociologists like Rogers wanted to know how technologies such as 2,4-D spray, antibiotics, and hybrid seed corn were transforming American agriculture. Despite heavy marketing by salesmen and widespread outreach from state agricultural agents, Rogers found that it had taken five years for the majority of the Collins community to adopt 2,4-D after its first use by a local farmer.

Rogers, however, was after a bigger target than farmers' use of pesticides. Whether farming technologies or homemaking fashions, he wanted to know how new ideas emerged, how these "innovations" traveled, and how they could be spread faster. Rogers had in mind a universal theory of how ideas spread. In 1962, at the age of thirty, he published his findings in *The Diffusion of Innovations*, a book that became the second most cited social science study of all time.[3] The adoption of new ideas, Rogers argued, changed entire societies through a cascading series of individual choices. What's more, he found that people's individual attitudes toward innovations had a major impact on shaping society. Those who were the most parochial and the most cautious were the least likely to contribute to social change. Conversely, those who were the most cosmopolitan and the most venturesome played an outsized role. Whether in farming, medicine, or scientific research, these "innovators" were the leading edge of societal change.

Innovation was not a novel idea when Rogers stepped onto the Iowa cornfield. The term had been in popular speech at least since the Protestant Reformation when the Catholic Church wielded it as an epithet against heretics.[4] Rogers was also not unique in claiming authority over how innovation worked. At the turn of the twentieth century, the French sociologist and criminologist Gabriel Tarde had used the concept to describe society as an ever-changing network of people, ideas, and technologies.[5] In the 1930s, the American sociologist William Fielding Ogburn likewise identified society's inability to keep up with new technologies, its *culture lag*, as a pressing national concern.[6] Still, prior to World War II, innovation was not a household word. It did not play a central role in American thought and culture. It was a term for specialists inside bureaucratic organizations and applied fields such as rural sociology.

When Rogers conducted his fieldwork, however, interest in innovation's nature had exploded. His literature review documented over five hundred works that used the term published since 1940 in sociology and anthropology alone. While Rogers was interviewing farmers, Bell Laboratories vice president Jack A. Morton promoted a system of "organized creativity" as the source of the company's world-changing technologies. At Harvard

Business School (HBS), the charismatic war hero and venture capitalist Georges Doriot taught students that innovation was a novel form of scientific entrepreneurship. Across the Charles River, MIT economist Robert Solow developed a Nobel Prize–winning theory of technology as the driver of economic growth. And Carnegie Corporation president John Gardner claimed that innovation could renew a nation confronting its moral decay.

This chapter investigates the multidisciplinary roots of innovation culture. It explores how different communities of experts developed theories of innovation to solve local and disciplinary problems that coalesced into a shared belief that innovation held the answers to the nation's future. It documents how, in the process of building this consensus vision, Rogers and others became the first innovation experts. It explores what role the special class of people called "innovators" play in their theories. Finally, it traces how, like 2,4-D weed spray, their visions spread.

FOUR QUESTIONS INNOVATION ANSWERED

It is not difficult to understand why midcentury experts claimed to discover novel truths about how societies changed. "The gulf separating 1965 from 1943," the anthropologist Margaret Mead wrote at the time, "is as deep as the gulf that separated the men who became the builders of cities from Stone Age men."[7] Experts themselves had witnessed astonishing changes in their lifetimes from the use of atomic weapons to the rise of microelectronics and digital computing, the birth control pill, the extension of automation to hundreds of industries, commercial jet travel, new consumer goods, and astonishing rates of economic growth. The gulf that Mead described had world-changing benefits but also profound disruptions. Experts worried that scientific and technological advances were unequally distributed and that the nation risked splitting into haves and have-nots. To hear these experts tell it, they were developing innovation models to accelerate the creation and uptake of technologies, foster growth, and remedy the undesirable impacts of a scientific age.

Historians of the Cold War, in contrast, argue that these experts used their theories of innovation for primarily ideological ends. For example, a dominant postwar theory of innovation, which historians later labeled the "linear model," posited that scientists made discoveries, engineers turned them into products, and corporations distributed them through society.

This linear model conveniently justified a military–industrial–academic complex in which the federal government was the principal financier and both academic scientists and large corporations were prime beneficiaries.[8]

Both naturalistic and interest-based explanations for the rise of innovation culture contain important truths and limitations. The reality of geopolitical shifts and technological change insists that we take innovation experts and their theories seriously. Yet the alignment of experts with vested interests reveals that their theories were inextricable from political intentions.

A more fruitful approach for understanding innovation culture's emergence in the mid-twentieth-century United States is to investigate what questions experts believed that innovation answered. This pragmatic interpretation brings to light the multidisciplinary origins of innovation both as a scientific explanation and as a cultural value.[9] Because different communities confronted distinct questions, the meaning of innovation differed depending upon who was asking. From 1945 to 1960, distinct clusters of experts independently called upon innovation to address four of modernity's biggest challenges. These experts developed theories of innovation in search of the underlying mechanisms for creating new technologies in a bureaucratic scientific age, accelerating the spread of ideas in a mass society, generating economic prosperity, and improving civic life.

HOW ARE NEW THINGS CREATED?

A decade before Rogers's work on 2,4-D adoption, World War II appeared to settle, once-and-for-all, the tension between independent inventors, previously championed as the engine of American ingenuity, and the laboratories of large corporations responsible for the military and consumer breakthroughs of twentieth-century life. Throughout the first half of the twentieth century, the locus of new technologies shifted from independents and small firms to corporate research and development. Corporate laboratories benefited from economies of scale that attracted the best university recruits, capitalized on interdisciplinary collaboration, and benefited from legal and financial power to control patent rights.[10] During the war, their executives had occupied key government leadership roles that helped their companies grow as the federal government became their largest customer and the nation's largest research manager.

Changes in the postwar R&D landscape, however, raised several challenges for technology corporations. How could companies sustain creative genius in environments criticized for producing organization men? Could corporations compete with universities for PhD scientists who valued basic research over practical engineering? How could companies plan and survive for system-transforming, rather than merely incremental, technological advances? In the late 1950s, the concept of innovation provided an integrative explanation for high-tech research and development.

A single device, the transistor, provided the model for what innovation was and how it worked. In the late 1940s, the team of physicists William Shockley, John Bardeen, and Walter Brattain created their Nobel Prize–winning discovery at Bell Laboratories. The transistor solved a critical problem of electronics—how to overcome the unreliability, scale, and fragility of vacuum tubes. Bell Laboratories, the twentieth century's preeminent corporate research facility, translated a decade of research on the principles of solid-state physics into a viable device that became the building block of digital society. The transistor would come to symbolize the entire economy and culture of innovation, the seed of Silicon Valley and, ironically, Bell Laboratories' decline.

Today, the transistor has mythical status in the history of innovation, yet the term "innovation" was not a part of Bell Laboratories' lexicon until after its creation.[11] The concept's first use came when the organization took a rare look backward. On the tenth anniversary of the transistor's invention, Bell Laboratories held a symposium to boost morale and to remind the national press of the lab's importance. Just two years prior, Bell had basked in the publicity of its Nobel Prize–winning discovery. But in the interim, Bardeen had left to teach at the University of Illinois. More disconcerting, Shockley departed for California, where he attempted to bring transistors to market on his own. To make matters worse, Bell Labs' parent company, AT&T, risked losing its monopoly when the Department of Justice forced the company to grant nonexclusive licenses on the transistor and other patents.[12]

Despite Bell Labs' setbacks, Jack Morton, the research manager who oversaw the transistor's development, was feeling proud and defiant. Morton had joined the company in 1936 as a young PhD student, and, like many engineers, transitioned into management. Between 1948 and 1955, Morton turned the bench-top model of the transistor into a commercial product.

Along the way, he made decisive bets, including choosing silicon as the base material for the microelectronics industry. As a result, more than 47 million transistors would be sold in 1958 alone. After two decades of service behind the scenes, Morton had been promoted to Executive Vice President. At the anniversary celebration, he boasted about the unheralded "innovation" process behind a device that would "have greater impact on society than the nuclear extension of man's muscle."[13] In a 3,000-page technical history prepared for the event, Morton wrote that the transistor had been the result of a systems-engineering approach that combined the "anticipation" of future needs, the flow and feedback of "new concepts, theories and materials," the design theory of the device, and an understanding of the broader environment in which it functioned.[14]

Shortly after Morton introduced innovation as a term of the art inside Bell Labs, another veteran executive, John Pierce, called on "innovation" to popularize the lab's approach outward to the public. In *Scientific American*, Pierce argued that technological innovations were "as important as language, art and science in distinguishing man from beast." Humans had been innovators since the use of stone tools, Pierce wrote, but the post–World War II convergence of science and technology in large organizations qualitatively changed the nature of innovation. He argued that modern innovation required advances in basic sciences, collaborative teams "working on matters of great personal, as well as mutual, interest," large organizations with clear goals, and, most of all, "response to a need." The transistor again was the definitive case, marrying "a new field of science" with "creative technology."[15] Innovation—the root of social change—flowed from the technoscientific corporation and its research managers like Morton.

While Bell Laboratories was defining innovation as the purview of large-scale R&D, independent inventors had also begun using the concept to defend their declining relevance. In the late nineteenth century, inventors were heralded as heroes of American ingenuity and enterprise. However, the rise of corporate R&D and the growing scale and scientific complexity of technological systems in the first decades of the twentieth century had left independent inventors struggling and demoralized.[16]

During World War II, independent inventors made a comeback when they banded together in a new National Inventors Council (NIC) to produce critical technologies for the war, including Samuel Ruben's miniaturized dry-cell battery. These wartime successes gave independent inventors a small foothold in the military–industrial complex. The NIC advocated for inventors in the face of dwindling public attention and financial viability, pointing to the independents' critical but overlooked contributions.[17]

A few successful independents with similar skills as their corporate peers also found niches in the burgeoning electronics industry. Jacob Rabinow, for example, a former electrical engineer at the National Bureau of Standards, left the government in 1954 to develop and bring to market magnetic storage devices for high-performance computing. These science-minded independents pointed to categorical differences between conception and application. "Invention" was about the creation of novel ideas and proof-of-concept, whereas "innovation" was about bringing ideas to market. This distinction created the impression of linear stages in R&D and identified different kinds of people and organizations with each stage.[18]

In the 1950s, a new type of scientific inventor–entrepreneur also emerged between the corporate research laboratory and the independent inventor. Clustered around Stanford and MIT, these were the entrepreneurial scientists behind the technology spin-offs that threatened AT&T's monopoly. Shockley's partnership with Beckman Instruments to beat AT&T at its own game led to the creation of Fairchild Semiconductor and a slew of other Bay Area companies. In 1957, MIT engineers Kenneth Olsen and Harlan Anderson likewise founded Digital Equipment Corporation (DEC) to manufacture minicomputers and challenge IBM's dominance. But such start-ups were not limited only to Cambridge and Palo Alto. The fast-growing Texas Instruments, which formed in 1951 as a spinoff to an oil industry R&D company, began manufacturing transistors with a patent license from AT&T. These companies were transforming what it meant to be an engineer, entrepreneur, inventor, industrial scientist, research manager, or scientist. The R&D landscape seemed to call for new understandings of technological development and a new vocabulary to describe it.

The common factor behind the successes of corporate laboratories, scientific independents, and microelectronics entrepreneurs was government contracts, a point not lost on the consultants who profited from the result.[19] In the early 1960s, management consultants gave theoretical grounding to the problems of scientific managers and entrepreneurs. These experts sought to understand the barriers to bringing new technologies into the world and "innovation" became a powerful explanation for identifying and overcoming those barriers.

In 1963, Donald Schön, a young consultant at Boston's Arthur D. Little, offered a theory of technological breakthroughs that focused on a special class of people he called "champions for radical new inventions." Writing in the *Harvard Business Review*, Schön synthesized work that he completed under contract for the NIC that examined how the relationship between the military and business simultaneously drove demand for innovation but limited its creation in practice. Both sectors claimed to want radical innovation: the military demanded new technologies to compete with the Soviets, while businesses required new products to stay ahead of competitors. However, both sectors were "caught in the gap" between a desire for "deliberate and systematic methods of innovation" on the one hand and "uncertainty and risk" on the other. They thus promoted innovation as a goal but resisted it in practice.

Schön applied the strengths of independent inventors to the context of large organizations. He noted that all new ideas encountered "sharp resistance" and that the ones that succeeded were almost always those that had a single champion. "Where radical innovation is concerned," Schön wrote, "the new idea either finds a champion *or dies*." These champions risked their careers in pursuit of their ideas. They needed to have "considerable power and prestige" and to know how to navigate informal political channels from within; but these very qualities also put them at odds with management. Schön concluded that "the most important aspect [of R&D] is not formal organizational procedures but the social tangle."[20] In other words, technological innovation was a social and cultural problem.

HOW DO IDEAS SPREAD?

While R&D managers looked for new ways of turning ideas into practical technologies, social scientists independently developed theories of innovation to describe the societal impact of those ideas and technologies. For anthropologists, "innovation" took on a broad meaning for the transit of ideas and behaviors. University of Oregon professor H. G. Barnett identified innovation as the defining source of human life, a universal process that linked aboriginal childrearing to nineteenth-century religious cults.[21] Building on Ogburn's influential culture lag model, academic sociologists emphasized technological innovations in particular and sounded the alarm that innovations from television to nuclear weapons were spreading faster than society could handle.[22] Meanwhile, MIT historian Elting E. Morison studied the adoption and resistance to individual technologies such as navy weapons to conclude that openness to innovation was a distinctly American value.[23]

It was the discipline of rural sociology, however, that elevated innovation studies to a formal research field. A quintessential praxis discipline, rural sociology grew out of efforts to address the gap between urban and rural life. In the 1910s, the United States Department of Agriculture (USDA) built the Cooperative Extension Service to distribute knowledge and technology to rural communities, while land-grant colleges turned a generation of farmers' sons into social engineers.[24] Rural sociologists published their findings in respectable academic journals, but they reached their largest audience through the extension stations of land-grant universities.[25]

Rural sociology's central role in defining innovation stemmed from the invention and spread of hybrid seed corn. Few twentieth-century innovations were as impactful. In the early 1900s, agronomists applied Mendelian genetics to crossbreed drought- and disease-resistant strains of corn. From 1933 to 1939, hybrid seed corn usage in the United States skyrocketed by 6,000 percent, from 40,000 acres to 24 million acres of crop. Growth was especially dramatic in the Midwest. By 1939, three-quarters of all corn grown in Iowa came from hybrid seed.[26] This stunning change occurred amid the Great Depression when capital and credit were limited. Farmers, however, were willing to experiment because they were desperate for a lifeline. The USDA, moreover, worked with commercial producers to distribute new crop strains. Research-oriented farmers such as Henry A. Wallace, who rose from teenage farmer to vice president of the United States, also contributed to hybrid seed's success.[27]

Hybrid seed corn was an ideal research topic for rural sociologists. Funding was plentiful because the agencies and corporations that developed hybrid seed were financing academic studies of its adoption. The data was also rich. Because adoption was so swift, farmers could describe their prior growing practices, why they chose to adopt the innovation, and exactly when they made their choices. Finally, hybrid seed's impact was undeniable, showcasing the role of modern agriculture in national progress.[28]

In 1941, two decades prior to Rogers's work, Iowa State graduate student Neal Gross developed the basis of the diffusion model of innovation. Gross conducted 345 interviews with farmers in a period of just two months. He

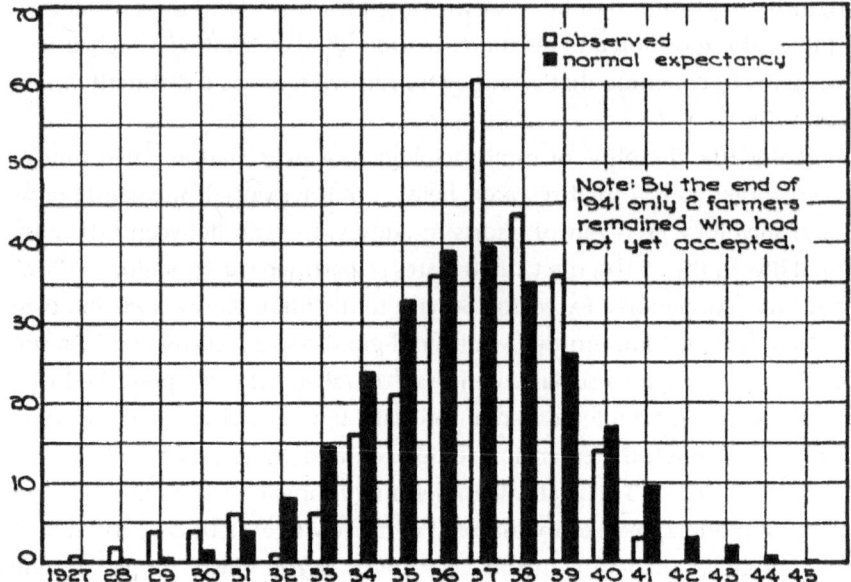

FIGURE 2.1
The normal distribution of innovation adoption discovered by Neal Gross and Bryce Ryan. Courtesy of Rural Sociology Society.

and his advisor, Bryce Ryan, found a lag between the time that farmers first heard of hybrid seed and when they adopted it. Adoption seemed to follow a normal distribution from the first farmers to experiment with it to the last farmers to accept the change. However, the pair doubted the universality of their findings and warned against modeling similar diffusion curves of "fads and fashions."[29]

Despite Gross and Ryan's warnings about generalizing from their corn studies, diffusion studies rapidly diffused. Rural sociologists extended the paradigm from Midwestern farms to Asian rice paddies as they contributed to Cold War modernization projects.[30] A key vector in this expansion was the 1955 report *How Farm People Accept New Ideas*, which grew from an initial run of 25 copies to 80,000 copies in seven years. The report explained in simple language how individuals took up innovations and how they spread throughout communities.[31] A variety of fields, including mass communication, teacher education, industrial relations, and medical sociology, adopted the diffusion model. In one influential study of the pharmaceutical industry, for example, a team of scholars funded by the Charles Pfizer

Company explored how an anonymized "miracle drug" grew from limited use to almost universal prescription by doctors.[32]

Diffusion theorists stressed the scientific nature of social change. They utilized statistical survey studies and projected disinterested objectivity. The metaphor of diffusion was itself a bid toward the physical sciences. But diffusion experts were defined by their interests. Their work was funded by the federal government, land-grant universities, and corporate manufacturers. Their goals were enhanced yield at a faster rate, communication strategies that would move product, and techniques that would reduce user resistance. Diffusionists thus shared common purpose with economists—a desire to turn innovation into money.

WHERE DOES WEALTH COME FROM?

For almost two centuries, modern economics rested on the assumption that wealth comes from a combination of labor and capital investment. By the mid-1960s, however, economists increasingly pointed to *technical change* and then *innovation* as the driving force of growth. Meanwhile, business schools helped to invent and promote new methods for funding high-tech ventures. As was the case among engineers and social scientists, these economic interpretations of innovation cut across schools of theoretical and applied thought. Where technical and social theories of innovation emerged in suburban New Jersey and rural Iowa, the rise of innovation inside economics was intensely concentrated within a two-mile radius centered in Cambridge, Massachusetts.

Economic theories of innovation were rooted in efforts to explain what caused the Great Depression and how to respond to its shocks. Harvard's Joseph Schumpeter argued that the boom and bust of the economy were determined by three major forces: external factors such as wars and natural disasters, growth factors such as increases in population, and novel transformations that didn't fit these predictable steps; in other words, innovation. "Innovation," Schumpeter wrote in 1935, generated "historic and irreversible change in the way of doing things."[33] The root of innovation's irreversible changes was not simply technology. Rather the *entrepreneur* was a crucial pacesetter of change. These "New Men" were irrational, persistent, and occasionally wildly successful at creating new industries even as they accelerated the death of existing ones. This process of *creative destruction* was capitalism's blessing and its curse.[34] Writing amid the dual rise of communism and fascism, Schumpeter concluded that free-market capitalism could result in recurring disasters if left unregulated by the state. However, attempting to eliminate creative destruction through government control would be just as devastating.[35]

Instead of a feared second depression, the United States experienced dramatic economic growth after World War II that led many economists to question what explained the transformation. Robert Solow, an economics prodigy at MIT, placed innovation front and center in his answer. Solow had entered Harvard in 1940 at the age of sixteen, a first-generation student of working-class Jewish parents. After the war, he returned to Harvard for his PhD before taking a position with MIT at just twenty-five years old. In 1957, Solow mathematically "proved" the source of national productivity. He began by showing that the contribution of capital (K) and labor (L) to output (Q) in the American economy had not fundamentally changed. If capital and labor did not account for the growth, what did? The answer was "technical change" which performed a similar role to Schumpeter's innovation. Technical change was not just technology, but also "slowdowns, speed-ups, improvements in the education of the labor force, and all sorts of things." Using the time series $Q = F (K, L; t)$ with data from 1909 to 1949, Solow derived rates of technical change from fluctuations in output that could not be explained by labor and capital. These year-to-year changes appeared random. However, the cumulative outcome was a dramatic upward curve from which Solow concluded that nearly 90 percent of growth came from technical change.[36]

Solow's approach earned him a Nobel Prize and became a leading theory for economic growth, but initial reception to his production function was mixed among microeconomists who focused on the ground-level details of technology and business.[37] One of Solow's earliest critics, Zvi Griliches, approached the problem by way of hybrid seed corn. In contrast to Solow's abstract model, Griliches emphasized that the rate of technical change depended on local community dynamics, a perspective that he and colleagues had started to call the "economics of innovation."[38] MIT's W. Rupert Maclaurin, for example, had studied how new industrial sectors such as radio and fluorescent lighting had emerged out of the interplay between engineers, investors, and competing firms pursuing individual inventions. In 1953, Maclaurin posited a model of technical change that emphasized

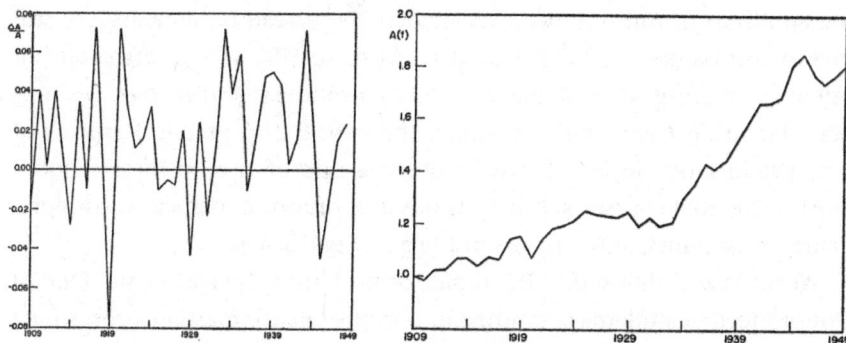

FIGURE 2.2

Technical change as the driver of the economy. Robert Solow shows how seemingly random shifts in annual rates of technical change (*left*) had an accelerating impact on growth (*right*). Courtesy of the MIT Press.

distinctions between "the propensity to develop pure science, . . . to invent, . . . to innovate, . . . to finance, and . . . to accept innovations," in which "the innovator" was "the most sensitive individual figure in the economy."[39]

In the early 1960s, the economic study of technological innovation became a unifying question of the field. The postwar rise of economic research centers that included the RAND Corporation and the National Bureau of Economic Research amplified the ideas of a group of young scholars—among them Kenneth Arrow, Edwin Mansfield, Richard R. Nelson, Jacob Schmookler, and Sydney Winter—who blended insights from macro- and microeconomic theories of innovation.[40]

The relationship between new technologies and economic growth, of course, was not just an academic problem, it was the central preoccupation of any technology-focused business. Bringing new technologies to market is as much of a financial challenge as it is a technical one. As we saw in the case of Bell Laboratories, the development and launch of new technological products underwent significant changes during and after World War II that required new managerial techniques but also new financial strategies.

At the Harvard Business School, the French émigré George Doriot helped identify these strategies in technology financing, theorized them, and trained generations of managers in their application. The son of a Peugeot automobile engineer, Doriot enrolled at HBS after World War I with the ambition of running his own factory. He instead became a successful investment banker. In 1926, Doriot returned to HBS as a young professor intent on sharing what he had learned as an industrial financier. His new Manufacturing Class, which blended theoretical and practical aspects of new product development, would become one of the business school's most influential courses, taken by thousands of future managers, entrepreneurs, consultants, and scholars in the ensuing decades.

World War II threw the HBS professor back into the real world. One of Doriot's former students appointed him to the Quartermaster Corps, where he rose to the rank of brigadier general. During the war, Doriot oversaw the research and development of dozens of new products from flight suits to Saran wrap. From his wartime position, Doriot witnessed the complexities of science-based technology and the barriers facing scientists who wanted to commercialize their ideas. In 1946, with the aid of MIT president Karl Compton and former US Senator Ralph Flanders, Doriot created the American Research and Development Corporation (ARD), a new kind of funding organization to assist in science-based commercialization.[41] ARD established the model of high-tech venture capital, priding itself on the scientific expertise of its small team who worked with university professors and corporate defectors who had been unable to secure conventional funding.[42]

For the venture capitalist, innovation was a process of giving "life and vitality to the combination of man and idea."[43] It was the venture capitalist's job to evaluate both the men and the ideas, and to use his expertise to help promising clients navigate the risks of the market. In practice, the techniques of early venture capital at ARD resulted from lessons learned from thousands of pitches and nearly one hundred investments. For Doriot, they included don't become too emotionally attached to an idea or a man, avoid overestimating an innovation's novelty, getting pricing wrong can mean failure, and never underestimate the competition.

In 1957, after assisting dozens of successful companies, ARD hit a home run. The firm invested $70,000 to help Olsen and Anderson's computer company DEC get started. From an old wool factory outside Boston, the start-up challenged the corporate giant IBM during the minicomputer revolution to become a multimillion-dollar business. DEC's success made ARD rich and made venture capital a recognizable industry, one that reinforced the lesson that innovation stemmed from experience rather than mathematical equations. ARD's seed funding for DEC also proved that expertise in innovation could be a highly lucrative means for building imagined futures.[44]

HOW CAN THE WORLD BE CHANGED?

Despite national fervor for technological change, many experts in the post-war era found modernity wanting. The devastation of World War II continued to shock. The rise of big corporations, big government, and big science seemed to sap human creativity and threaten individual autonomy. Technological prosperity improved the comfort of millions of Americans, but also fostered soullessness. The knowledge economy that generated these benefits required advanced education, threatening to widen inequality and disenfranchise workers. Minorities denied access to prosperity demanded basic rights and the correction of systemic injustices. The threat of Soviet communism compounded the problems of individuals in a mass society. The most acerbic critics declared the death of progress, while reformers asked how democratic participation could exist in a corporatist, industrial state.

Innovation offered an answer to these dilemmas of modernity for a heterodox group of social psychologists, business consultants, and bureaucrats. This social innovation approach also had origins in depression-era Iowa and later flourished at MIT; Washington, DC; and eventually inside corporate boardrooms. Collectively, their work gave larger social and ethical meaning to the concept of innovation.

At the University of Iowa, psychologist Kurt Lewin, a German-Jewish émigré, worked with his student Ronald Lippitt to explore how groups shaped individuals. Lewin was a pioneer of "action research," an approach premised on the belief that knowledge stems from cycles of planning, engagement, and observation in which experts were themselves participants. Rather than observe how change happened, action researchers initiated the changes.

Through the Iowa Child Welfare Research Station where Lewin was director, the pair worked with local schools to explore how children responded to different forms of social organization. To test their ideas, they set up three different youth clubs for kids to play games, make crafts, and go on adventures. One club had a strong autocratic leader who gave instructions on exactly what children should do. Another was laissez-faire, leaving

kids to their own devices. A third was democratic, with the leader facilitating collective decision making. Children flourished in the democratic
group, where they shared more and tolerated their peers. Yet when Lewin
and Lippitt switched kids in the democratic group to the authoritarian or
laissez-faire groups, they reverted to obedience or bullying in the former
and cynicism in the latter.[45]

Lewin and Lippitt built on these experiments to develop one of the most
influential tools of organizational change—the training group or T-group.
In 1946, Lewin and Lippitt moved to MIT at a new Center for Group Dynamics. Among the Center's first initiatives was a project to confront religious
and racial prejudice in the struggling city of Bridgeport, Connecticut. A
center team brought together schoolteachers, social workers, community
organizers, businesspeople, gang members, and real estate developers in
a T-group that utilized facilitated conversations and role-playing games.
The organizers claimed remarkable results in confronting misunderstand-
ings and generating collective community growth. With funding from the
National Education Association and the Office of Naval Research, Lewin
established the National Training Laboratories for Group Dynamics (NTL) to
train a generation of disciples. NTL also developed the T-group model into
lucrative management retreats to help companies confront organizational
stagnation.[46] In his 1958 book, *Dynamics of Planned Change*, Lippitt linked
this work to "forces toward innovation in human society."[47]

In the 1950s, a jack-of-all-trades named Warren Bennis used the language of social innovation to categorize theories, practices, and tools of
action research. Like other early innovation experts, Bennis's ideas developed along with his eclectic personal and professional path. He served as an
infantry officer in World War II when he was still a teenager. After the war,
he studied existential philosophy at Ohio's experimental Antioch College.
He then enrolled in a PhD in economics at MIT. In 1955, while a graduate
student, Bennis spent a life-changing two weeks at the "micro-utopia" of
the NTL. Participants were encouraged to shed societal norms in search
of authentic relationships and social practices, "people screamed, people
guffawed, people wept, people talked . . . emotions circulated between the
members of the group, creating some superior new social organism."[48] His
unorthodox dissertation built on the experience, investigating barriers to
teamwork in an interdisciplinary research institute he called the "Hub."
Among his arguments was that "innovators" were those who found creative

solutions to foster cooperative relationships inside systems that resisted change.[49]

In 1961, Bennis, then a member of MIT's Sloan School of Industrial Management, coedited *The Planning of Change*, an anthology on the applied behavioral sciences that portrayed social innovation as the act of guiding open-ended, human advancement. Collaboration was essential for innovation, he wrote, not only with respect to "scientific objectives" but also as "an ethical imperative."[50] Bennis then left MIT to apply his concepts as provost of the State University of New York at Buffalo as the university attempted to remake itself on his decentralized principles. Along the way, Bennis published reflections on his work that helped to establish the field of leadership studies.

John Gardner, an almost forgotten figure today, further elevated social innovation as a political and ethical philosophy. Gardner is best remembered for founding Common Cause in the 1970s, one of the first citizen-advocacy groups, responsible for campaign finance reforms and for lowering the voting age in the United States to eighteen. But prior to his voter advocacy work, Gardner was a fixture of the Washington policy elite and a leading advocate for institutional reform to unlock the innovative potential of every American.

Gardner grew up on the margins of poverty, raised by his widowed mother in rural Beverly Hills as it changed from an exurb to a home for movie stars. He was a talented student and athlete who became a champion swimmer at Stanford University. After graduate school in experimental psychology at Berkeley, Gardner then served as an intelligence officer during World War II, overseeing democratic propaganda in Latin America. After the war, he joined the Carnegie Corporation and swiftly rose to become its president.

For Gardner, innovation provided the answer for the dual enhancement of individual potential and social progress. In his influential 1961 treatise *Excellence: Can We Be Equal and Excellent Too?*, Gardner identified "talent" as a national priority but argued that squaring democracy with meritocracy demanded reform across American institutions.[51] Gardner laid out a vision for a pluralistic, tolerant, and open society in the book *Self-Renewal: The Individual and the Innovative Society*. He argued that selfishness, institutional stagnation, and moral apathy portended American decline. However, "reflection" and "renewal" could re-infuse moral purpose in a democratic society.

In 1965, Gardner had a mandate for his ideas when President Lyndon B. Johnson appointed him to lead Great Society programs at the United States Department of Health, Education, and Welfare. Among Gardner's core concepts was "continuous innovation," an approach now repeated by administrators and managers around the world. The "endless" process of innovation, he wrote, was a "burdensome responsibility," but it was essential because "a society is being continuously re-created, for good or ill."[52] Gardner also helped normalize the "public–private partnership" as an explicit strategy for civic policy. Unlike later interpretations of public–private partnerships aimed at reducing government services, Gardner believed that collaboration could create buy-in for and decrease resistance to federal Great Society interventions in education, public health, economic opportunity, and civil rights.[53]

The social innovation experts drew moral assumptions about innovation that were meant to be just as effective for school children, urban communities, corporate managers, and government crusaders. They critiqued structures that sought to preserve the meritocracy and argued that everyone could participate. In Gardner's view, practitioners of "plumbing" and "philosophy" alike could achieve excellence in an innovative society.[54]

BECOMING INNOVATION EXPERTS

By the early 1960s, practitioner–scholars in four broad domains were working to crack the secrets of innovation. None referred to themselves as innovation experts. Nor were most connected with each other in formal networks. As these champions for innovation achieved success, however, they began to share common cause.

Early innovation experts demonstrated similar personal qualities and career trajectories. They were itinerant travelers across the institutions of postwar American society. Most had found success at a young age, but between the margins of disciplines, organizations, and societal sectors. Their careers blurred boundaries between academia, consulting, government service, and organizational leadership. These experts also self-consciously perceived themselves as both insiders and outsiders. They were facilitators and managers, rather than inventors or scientists. Simultaneously thinkers and doers, they were proud of their ability to link theory and practice.

These plumber–philosophers defined themselves as "change agents," a role characterized by the ability to judge the desirable direction of social change and to intervene in the adoption or rejection of new ideas or technologies to make that change possible.[55] They developed textbooks, seminars, and retreats to help organizations diagnose their problems, assess their willingness and capacity for change, identify their goals, and achieve their objectives. Early innovation theorists, in other words, set a pattern for a new kind of protean problem-solver.

EVERETT ROGERS, IDEAL TYPE

To better understand how this diverse cast of practitioner–scholars fashioned themselves as experts for innovation as scientifically observable, reproducible, and desirable, we return to the career of Everett Rogers.

Rogers set the mold of the innovation expert. He formulated his ideas at a remarkably young age. He was stylish and cosmopolitan, with a penchant for Brooks Brothers suits and Porsche convertibles.[56] His work was self-consciously interdisciplinary. He was incredibly prolific, authoring 36 books and 176 peer-reviewed articles.[57] He attracted an astonishing number of disciples, training over 150 graduate students, and marrying two. Finally, he promoted rule-based systems that were easy to remember and universal in application.

Rogers knew what it was like to live in a fixed social system. He was raised in Carroll, Iowa, at the height of the Depression. His father, an engineering school dropout, was an electromechanical tinkerer but refused to adopt hybrid seed corn until the family crops failed.[58] His mother was a well-read farmer's wife. He was educated in a one-room county schoolhouse.

Rogers's future was on the family farm, until a scholarship at nearby Iowa State changed his life. Few students from Rogers's rural county attended college; when they did, like Rogers, they majored in agriculture and supported their education through the Reserve Officer Training Corps (ROTC). But Rogers stood out as a talented and curious student, augmenting his coursework on fertilizer chemistry with creative fiction and journalism.

Rogers graduated Iowa State during the Korean War. After a brief stint training Air Force recruits, he became the personal aide to General Donald Flickinger, the director of an experimental Human Factors research team that gathered engineers, psychologists, and doctors to tackle problems at the human–machine interface. As superiors worked to prevent ejection seats from injuring pilots, Rogers redesigned military pants to accommodate pilots' growing waistlines. The Human Factors unit was a hotbed of interdisciplinary research where Rogers recounted meeting future internet pioneer J. C. R. Licklider as well as many prominent science fiction authors of the time who had been recruited to aid in psychological warfare.[59]

Sparked by an interest in the social dimensions of technology, Rogers returned to Iowa State for his PhD. When he arrived, the Department of Rural Sociology already was a global center for innovation studies. His dissertation on 2,4-D weed spray added one more case to the already robust diffusion model.

As an assistant professor at Ohio State University, Rogers expanded his dissertation's literature review into his 1962 book *Diffusion of Innovations*.[60] Structured as an introductory textbook, *Diffusion of Innovations* was a scholarly intervention into multiple research communities. By standardizing innovation's terminology across sociological, economic, and psychological traditions, Rogers united parallel tracks of innovation studies. He also positioned change agents like himself as cross-disciplinary experts who could aid in technology's implementation and assess its impact.

Diffusion of Innovations reduced complex phenomena to a toolkit of "ideal types" that broke down the innovation process into categories that could be remembered on the fingers of one hand. There were five kinds of innovations, five stages of adoption, five categories of adopters, and four sources of information impacting their decisions.[61] Importantly, ideal types provided change agents with actionable generalizations. Innovation experts, for example, should begin by targeting opinion leaders, they shouldn't force innovations upon clients, and it might be more effective to instill "norms of innovativeness" in a community than to focus on any specific innovation.[62]

Ideal types were supposed to be value neutral, equally applicable to brothels and to religion.[63] Rogers quipped, for example, that "a new narcotic among drug addicts" was a successful innovation. But, in practice, ideal types defined innovation as a social good. The diffusion model drew sharp contrasts between tradition and modernity. Traditionalists relied on subsistence farming, were provincial, had a low level of literacy, did not make rational economic decisions, and lacked empathy. The modern "cultural type," in contrast, was "more innovative, more progressive, more developed, or more economically rational" and had an "ability to empathize and see oneself in the other fellow's shoes."[64] Modern social systems, moreover, embraced technology, valued science and education, were cosmopolitan, and open to the free exchange of ideas.

While *Diffusion of Innovations* stands as the most influential work of innovation expertise ever written, it was not an instant blockbuster. No national newspapers or magazines noticed it. Academic reviews were lukewarm. Anthropologists and historians lauded Rogers's interdisciplinary approach, but sociologists and economists criticized the book's theoretical paucity and challenged its ideal types.[65] *Diffusion of Innovations*, wrote one reviewer, was "no classic."[66] Nonetheless Rogers's book bundled and synthesized existing work into a general theory that accelerated the diffusion model's spread. Even as critics identified its weaknesses, land-grant universities, professional schools, government agencies, corporate laboratories, and global development initiatives took up Rogers's vocabulary. In the second edition of *Diffusion of Innovations*, published in 1971, Rogers documented a 300-percent increase in innovation studies.[67]

Diffusion of Innovations propelled Rogers from the parochialism of rural sociology into the cosmopolitan role of innovation expert. As Rogers climbed the academic ladder, he traded farmer conventions and Girl Scout fashion shows for invited professorships and international consulting.[68] In 1963, Rogers traveled to Colombia on a Fulbright fellowship to study the global diffusion of innovations. In 1964, he joined Michigan State University's new communications department, where he received a stream of grants from the United States Agency for International Development (USAID).[69] In 1973, he moved to the University of Michigan with a joint appointment in public health and journalism where he mapped the diffusion of family planning methods. Two years later, Rogers landed at Stanford University amid the microelectronics revolution. Over the next three decades, he applied his diffusion model to a widening array of problems including computers, HIV prevention, and the war on drugs.

Rogers came to be lauded as a foundational thinker in communications, business, marketing, public health, and information technology by synthesizing prevailing expertise on innovation into a unified vision. Fittingly, this globe-trotting academic identified the *innovator* as society's most important change agent.

INNOVATION PERSONIFIED

Nearly all theories of innovation in the postwar era emphasized innovation as a "people" process. But who were the people who drove innovation? What made them tick?

Rogers and other rural sociologists led the way in defining the human qualities of innovation. Their diffusion model identified the "innovativeness of individuals" as the time it took someone in a social system to adopt an innovation compared to peers. The model assumed that people rationally weighed the advantages and disadvantages of an innovation against their values and norms. A key tool for mapping innovativeness was the sociogram, which captured the networked flow of ideas. As individuals encountered a prospective innovation, they sought information about its merits; if they decided to try the innovation, they would experiment with it on a small scale and adopt or reject it, and themselves become communication vectors.[70]

Rural sociologists empirically linked the innovativeness of individuals to specific personality types. In 1952, while Rogers was measuring pilots' waistlines, University of Wisconsin professor Eugene Wilkening found that in a community of one hundred farmers, ten were "local leaders" to whom nearly all others turned for advice. To his surprise, these leaders tended to be conservative and were not a major source of change. Instead, he discovered a small group of "community innovators" had over a period of twenty years introduced nearly every new technology and process in the region. These innovators had a high socioeconomic status but were not necessarily respected by their peers.[71]

Rogers made adopter categories central to the laws of innovation in *Diffusion of Innovations*. He boiled down human responses to change to five types along a normal distribution. Each type displayed a dominant cultural value and associated personal qualities. Most people fell into early and late majorities. They were influenced by their peers and adopted new ideas or technologies only after others had validated them. A smaller percentage was more willing to experiment. These *early adopters* had the capital to take risks and the education to evaluate those risks. In contrast, a minority of the population, usually older and more isolated, took up innovations only as a last resort.[72] Earlier studies had identified this group as "late adopters," "traditionalists," or "diehards." Rogers called them "laggards." Only one group

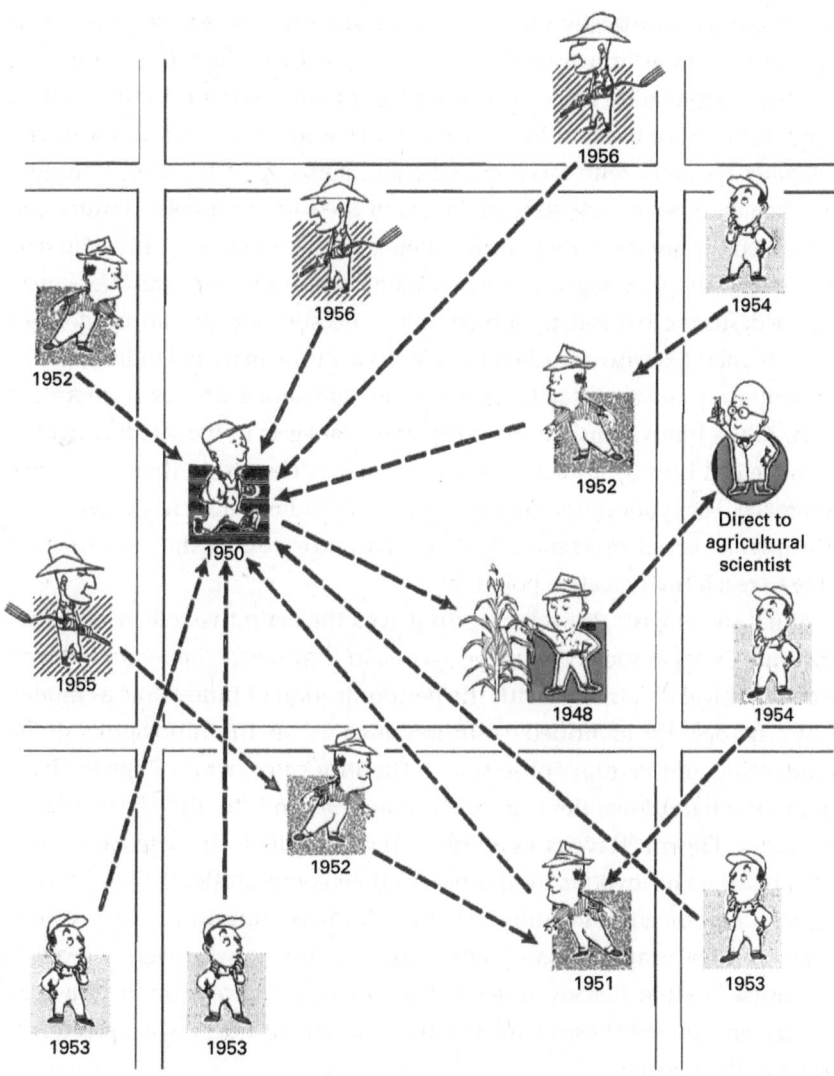

FIGURE 2.3
Rural sociologists used sociograms to map the people process of innovation, in this case documenting how farmers gained access to a new idea, beginning with an agricultural extension agent, and mapped across a one-mile grid.

stood out as statistically abnormal. Representing just 2.5 percent of the population, these "innovators" were "in step with a different drummer."[73]

This interpretation turned existing assumptions about innovators on their head. For most of Western history, innovators were unwelcome deviants and religious apostates who faced derision and persecution for their transgressive behavior. Social scientific observers in the early twentieth century corroborated this image as they turned their gaze to innovators. "Any who dare originate," the sociologist F. Stuart Chapin wrote in 1928, was "executed, lynched, stoned, hanged, or burned."[74] Two decades later, the anthropologist H. G. Barnett likewise described innovators as "truly marginal individuals."[75] In contrast, postwar research valorized the innovator as the most important actor in the innovation process. Corporate managers in the 1960s sought to emulate Bell Labs by attracting a stable of mavericks within their systematic approach. For economists, entrepreneurs were industrial society's New Men. Meanwhile, social psychologists described innovators as the citizens most able to reach their creative potential.

Rural sociologists, however, again played the definitive role in establishing innovators as society's change agents. In *Diffusion of Innovations*, Rogers wed statistical empiricism with the personification of innovators as modernity's heroes. He identified "venturesomeness" as the innovator's dominant value and central "obsession." The innovator, wrote Rogers, "must desire the hazardous, the rash, the *avant-garde*, and the risky." He and his colleague Eugene Havens established seven "salient characteristics" that distinguished innovators from others in their communities.[76] These characteristics reinforced innovation experts' self-image. Innovators were young, high in social status, cosmopolitan opinion leaders who were viewed as deviants.[77] While fiercely independent, innovators were often drawn to others who shared these characteristics "like circuit riders who spread new ideas as their gospel."[78]

Cultural difference and competition were baked into the persona of the innovator. Innovators were only identifiable against the norms of the social system from which they deviated. Moreover, while innovativeness existed on a quantitative spectrum, it was laden with judgment. To innovate was a moral virtue while to lag was a moral deficit. Innovators and laggards were both deviants, but only one was valuable.

Experts in business, management, and economics translated Rogers's interpretation of innovators into a wide variety of contexts. In one influential

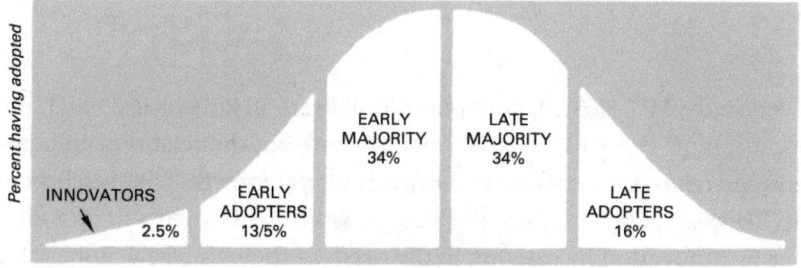

	Innovators	Early adopters	Early majority	Late majority	Laggards
Salient values	"Venturesome" willing to accept risks	"Respect" regarded by many others in the social system as a role-model	"Deliberate" willing to consider innovations only after peers have adopted	"Skeptical" overwhelming pressure from peers needed before adoption occurs	"Tradition" oriented to the past
Personal characteristics	Youngest age; highest social status; largest and most specialized operations; wealthy	High social status; large and specialized operations	Above average social status; average-sized operations	Below average social status; small operation; little specialization; small income	Little specialization; lowest social status; smallest operation; lowest income; oldest
Communication behavior	Closest contact with scientific information sources; interaction with other innovators; relatively greatest use of impersonal sources	Greatest contact with local change agents	Considerable contact with change agents and early adopters	Secure ideas from peers who are mainly late majority or early majority; less use of mass media	Neighbors, friends, and relatives with similar values are main information source
Social relationships	Some opinion leadership; very cosmopolite	Greatest opinion leadership of any category in most social systems; very localite	Some opinion leadership	Little opinion leadership	Very little opinion leadership; semi-isolates

FIGURE 2.4

Innovation classified and personified from a normal distribution to ideal types. This figure combines images from the North Central Rural Sociology Committee report *Adopters of New Farm Ideas* with Rogers's classic table of adopter categories with columns and rows reoriented to show the linkages between visual and textual representation. For the original, see table 6–4 in Rogers, *Diffusion of Innovations*, p. 185.

1968 study, MIT Sloan School professor Edward B. Roberts used MIT and its local spinoff companies as a model for studying the relationship between entrepreneurial innovators and organizational growth.[79] Research managers likewise sought to create the best working environments to promote scientific innovation by catering to the needs of their "internal" innovators.[80] Marketing experts similarly identified innovators as vital to the success of new products, speculating how to reach the "superinnovator."[81]

As innovation experts established the aggregate qualities of innovators, they rarely dwelled on individuals. A variety of popularizers, however, linked expert interpretations to the nation's past and desired future. Profiles of innovators appeared sporadically in popular culture during the first half of the twentieth century, almost without exception emphasizing the arts and humanities. They focused on iconoclastic artists and writers, such as Pablo Picasso, Igor Stravinsky, and Walt Whitman. Virtually all early twentieth-century profiles of innovators, moreover, were posthumous celebrations.

After World War II, the persona of the innovator shifted from prominent artists to scientists. In celebration of Albert Einstein's seventieth birthday, for example, *Scientific American* declared the physicist "a great independent innovator" rather than a gray-haired "ivory tower scholar."[82] Efforts to associate innovation with creative genius in the sciences built on associations with the arts, emphasizing the erasure of barriers between disciplines. In 1961, for instance, one of the earliest profiles of Leonardo da Vinci as an innovator stressed this blend of art, science, and technology to encourage similar patterns for present and future scientists.[83]

Business journalists also helped make the innovator a desirable social type. Magazines profiled executives to identify the path to "Being an Innovator" even as they lamented that corporations hired organization men.[84] The 1960 *Fortune* article "The Cultural Innovators," meanwhile connected corporate innovation to New York's experimental art world by identifying its gallery owners and promoters as "avant garde entrepreneurs."[85]

The elevation of innovation as a national virtue, however, came in the realm of technology. Popularizers emulated earlier portrayals of historic inventors but emphasized the innovator's ability to turn ideas into wealth and impact. For example, in *Edison: A Biography*, Pulitzer Prize–winning author Matthew Josephson sought to "remove the veils of myth" that cast Thomas Edison as a lone inventor. Tracing Edison's rise from entrepreneur to executive, the bestselling biography instead portrayed him as one of the first corporate innovators. "Today's giant research laboratories," Josephson wrote, "stand everywhere as monuments to Edison's innovative spirit."[86]

By the mid-1960s, popularizers combined this range of traits in the person of the innovator. The handsome coffee table book *The Fifty Great Pioneers of American Industry*, for example, wove together biographies of inventor–entrepreneurs such as Henry Ford, but also physicist Enrico Fermi, financier A.P. Giannini, union leader Samuel Gompers, and beauty industry pioneer Elizabeth Arden. It used these profiles to map "several fundamental characteristics" that these diverse Americans had in common. "They were not afraid to be innovators. They had no reverence for tradition. . . . And, above all, they simply could not be discouraged; they were almost incredibly persistent. Failure the first time or, for that matter, the third or even the fifth, was not a deterrent, but a challenge. . . . They were not content to prove something *could* be done. They were never satisfied until it *was* being done, practically and successfully, in the practical world."[87] Innovators, in short, took risks, made an impact at a young age, worked within a market context, and insisted on lifelong growth.

Once the qualities of innovators had been defined, seemingly overnight everyone wanted to locate, recruit, and cultivate them. A survey of job advertisements in major newspapers from 1945 to 1960 turns up just ten mentions of the trait. Between 1965 and 1969, however, ads seeking innovators skyrocketed to over one thousand.[88] Job seekers followed suit by adopting the identity as their core skill. In 1960, one intrepid engineer announced with bold all-caps his qualifications as an INNOVATOR who could "make advanced technology pay through sound commercial development."

Educators also began to explore how to identify and cultivate innovative talent. After Sputnik, the nation invested significant efforts in primary and secondary education to increase scientific manpower for the arms race with the Soviet Union. By the mid-1960s, however, the emphasis on standardization and testing shifted toward creativity-based approaches among progressive educators. Katherine E. Hill, a science education professor at New York University, argued, for example, that "there can be no doubt that children are innovators—producers of change; introducers of the new and untried." According to Hill, it was an important goal of educators to optimize the innovative behavior of students. Instead of being punished, "nonconformers" and "disruptors" should be channeled into creative pursuits. Moreover,

INNOVATOR

Highly qualified executive chemical engineer (M.I.T.) will direct technical activities of small or medium-size company. Establish realistic goals for process and product improvement. Provide feedback between general management and research and development. Apply research-by-contract to meet the needs for continued company growth. Make advanced technology pay through sound commercial development. Fifteen years proven ability in chemicals, minerals, and refractory metals.

Y 7631 Times

FIGURE 2.5
Innovator for hire. In the 1960s, "innovator" began to appear as a desirable professional trait. "Display Ad: Innovator," *New York Times*, January 11, 1960, 108.

the most innovative students were not necessarily those who scored highest on IQ tests. Educators therefore should "recognize all children as innovators" and accordingly restructure education away from standardized goals and metrics.[89] From the school room to the boardroom, America had a newfound demand for innovators.

INNOVATORS WANTED

In the first two decades after World War II, a nebulous but distinct vision of innovation as a mechanism for social change emerged out of diverse fields of practical and scholarly research. Innovation was to provide the means for distributing the fruits of corporate, government, and academic partnerships throughout rural and urban communities. These societal goals coincided with the more parochial goals. Innovation experts were defining new lines of research and asserting new roles as academic, corporate, and political consultants. The language and theories of innovation gave them a unique currency and a new identity.

Expert theories about innovation in the mid-1960s, however, contained a series of unanswered questions. If innovation stemmed from the ineffable creativity of individuals, how could it be routinized in organizations? Was innovation the purview of the free market or of state-directed progressivism? How could experts accelerate its uptake? *Should they* do so, as even champions like Rogers argued that people tended to adopt innovations as a "panacea"?[90]

By identifying *innovators* as agents of change, these experts brought the innovation process down to the human scale. This personification of innovation helped popularize their theories.

The nascent demand for innovators, however, revealed disagreements among experts. For one faction, innovators were a rare breed who accounted only for 2.5 percent of the population. Another faction argued that "every individual is basically innovative," and that a focus on "rare and brilliant individuals" gave a "distorted perspective of human ingenuity."[91] Experts also refused to limit innovation to science and technology. Gardner, for example, asserted that those who contributed "a new way of doing things" were just as important, if not more so, than those who made new things.[92] With proper training and support, he claimed, *everyone* had the capacity to innovate.

The emergence of innovation as a political goal posed another challenge, one that spoke to innovation experts' own identity and purpose. Innovation expertise seemed by definition to imply a preference for change. How could innovation experts balance their claims to objectivity with the transformational promises of their ideas?

Finally, asking what innovators were like begged a practical question: How might these valuable deviants be harnessed?

3   A NATION OF INNOVATORS

Invention and innovation lie at the heart of the process by which America has grown and renewed itself.
—United States Department of Commerce, 1967

On September 15, 1972, Big Bird made an unlikely friend in Richard Nixon. At a gala ceremony in the East Room of the White House, *Sesame Street* founders Joan Ganz Cooney and Lloyd Morrisett posed with the president and five other winners of the United States' first national prize for innovators. Sharing the stage with Willem Kolff, inventor of the artificial kidney, and Harold Rosen, whose geostationary satellite revolutionized world communication, Cooney and Morrisett received $50,000 to "encourage needed innovation in key areas of public concern."[1]

A press release described Sesame Street as a "classroom without walls, nationwide and infinitely expandable, capable of reaching into ghetto neighborhoods and remote rural outposts," improving the lives of preschool children "everywhere." Reaching ten million preschoolers in fifty countries at a cost of just one cent per child per hour, Sesame Street was the most "thoroughly researched" experiment in the history of television, a model of the "continuing flow of new and useful development; based on solid science."[2] Cooney, Morrisett, and the other winners showcased how creative Americans were applying science and technology in novel ways to solve society's greatest challenges.

Except it never happened. Despite Nixon's announcement of a Presidential Prize for Innovation in a special message to Congress, the receipt of nearly five hundred nominations, a selection committee representing seven federal agencies, an advisory board of industrialists, the selection of winners, and the drafting of congratulatory speeches, the award unceremoniously vanished.

This chapter uses the "Innovation Prize caper," as *Science* magazine dubbed the ill-fated award, to explore the origins of a central but underappreciated truth about the history of innovation culture.[3] Although often associated with entrepreneurial mavericks, laissez-faire economic models, and antigovernment ideology, "innovation" is in practice and concept a creation of the federal government. The government not only funds billions of dollars in basic research that fosters innovation, but it is also a leading innovation evangelist.[4] Every presidential administration has its "innovation policy." Bureaucrats in nearly every federal agency seek to build and maintain systems that link scientific research and development with regional growth, international competitiveness, and the training of innovative citizens.[5] It was not always so.

The Presidential Prize for Innovation is a window into how the government came to champion innovation decades before it was commonly recognized. It also explains what Sesame Street and kidney dialysis had to do with it. The contest came at a critical point in efforts to harness the new force of innovation in the nation's service. Unfolding during a moment of national turmoil, the prize was at the center of federal efforts to combat a litany of challenges from declining economic growth to environmental protection. The prize brought together framers of the nation's science policy, captains of industry, and the president of the United States. At the core of the government's efforts to identify national heroes, however, was a small group of bureaucrats.

The story of the abandoned Presidential Prize for Innovation is the story of the successes and failures of the federal government's first innovation evangelists. Between 1960 and 1974, a cadre of experts within the federal bureaucracy initiated a decisive shift in American beliefs about and policies toward innovation.[6] These experts were almost uniformly young, and had varied backgrounds in engineering, science, law, and philosophy.

Like Nixon's prize, these bureaucrats' contributions remain unknown despite an outsized influence in helping to make innovation a civic ideal. To achieve their vision of an innovative nation, they worked across democratic and republican administrations. They convinced politicians of innovation's value, fostered collaboration across agencies, and created policy that remade relationships between government, private industry, and the American public. They were as much interested in the moral value of innovation as they were in its economic effects. They portrayed innovation as a defining feature of the American character and sought to use the levers of government to imbue individuals, organizations, and society at large with an innovation ethic.

This chapter, then, focuses on two sets of innovators. First, the backstage *bureaucratic innovators* who translated innovation into a national vision. For these innovation experts, the Presidential Prize was to be the public face of a decade of infrastructural work cultivating new bureaucrats, government offices, public–private partnerships, and legislation. And second, the heroic Americans who the bureaucratic innovators placed front stage—the scientists, engineers, inventors, medical entrepreneurs, and children's television creators—that government policy aided in building a harmonious innovation nation. And yet a caper suggests clandestine plans gone awry. The story of how the federal government became an innovation evangelist is neither a triumphal celebration nor a well-coordinated and nefarious scheme. It is an account of fits, starts, and ideological ambiguity.

BUREAUCRATIC INNOVATORS

From the earliest days of the republic, the United States government has sought to foster what is now commonly understood as innovation. The Constitution established a Patent Office to grant inventors the rights to their work. In 1791, founders Alexander Hamilton and Tench Coxe penned their "Report on Manufactures" calling for tariffs to protect American industries and subsidies to build roads and canals to stimulate new manufacturing businesses. Congress likewise has long utilized the tax code to incentivize specific sectors of economic growth and social impact. So too has the government invested in research and education as a lever of social change, notably through the Morrill Land Grant Act of 1862, which established public universities charged with using the "practical arts" to improve communities and educate the industrial classes.

But it was during and after World War II that the government began to pursue innovation as a formal national goal. The nascent theories of innovation described in the prior chapter arose in concert with the government's expanded role as the leading financier and manager of scientific research and development. Vannevar Bush's *Science, the Endless Frontier*, which proposed the creation of the National Science Foundation (NSF), epitomized the postwar faith in the fruits of basic research. Bush's favored terms were *scientific discovery* and *technology*, rather than *innovation*, and the basic scientist was his leading agent of change. Bush's model, however, would come to be known as the "linear model of innovation," whereby innovations occur through consecutive stages of basic science, applied research, technology development, and diffusion.[7]

In the early 1960s, the federal government began to develop alternatives to this linear model. Experts and politicians concerned with the domestic economy sought to extend innovation to a broader range of agents and institutions. In the interplay of these challengers, innovation became a policy concept that encompassed projects as diverse as kidney dialysis and children's television.

The federal government hired its first self-styled innovation expert during the Kennedy administration. In 1962, J. Herbert Hollomon, General Electric's top research director, joined the Department of Commerce as its first ever Assistant Secretary for Science and Technology. Unlike the Cold War physicists who advised the government on super weapons from their positions in universities and national laboratories, Hollomon's heart was in industry. The forty-two-year-old metallurgist had earned his bachelor's and PhD from MIT while serving the war effort at the Watertown Arsenal. He then spent sixteen years rising through GE's ranks. He was a prolific writer, a visionary, a recruiter of talented misfits, and a polarizing reformer.

The Department of Commerce does not feature prominently in histories of postwar science and technology. Yet, since its founding in 1913, Commerce had always performed important technological functions that included the National Bureau of Standards (NBS), the Weather Service, and the Patent Office. By the late 1950s, however, the bulk of federal funding for science and technology went to the space and arms races with the Soviet Union. These initiatives were guided by the Department of Defense (DOD) and new agencies with scientific missions, including the Atomic Energy Commission (AEC), NSF, and NASA. Hollomon's appointment was a concerted effort to remake Commerce as a space age home for civilian technology.[8] He envisioned Commerce as a nimble research-based clearinghouse

FIGURE 3.1
J. Herbert Hollomon. Research Information Services General Electric Research Laboratory. Courtesy of MIT Museum.

and coordinator that could make the government an accelerator of American industry.[9]

More than any other individual, Hollomon helped to integrate "innovation" into the government lexicon.[10] He described innovation as the "key to a vibrant, competitive domestic economy and the best bet for the prevalence of American enterprise in the markets of the world."[11] He argued that the massive flow of resources into research and development had not produced effective results for everyday Americans because over two-thirds of the billions of dollars spent on R&D went to space, defense, and atomic energy.[12]

To implement his innovation mission, Hollomon recruited a team of young industry experts to the federal ranks. Two, in particular, shaped government attitudes about innovation.

The first, Donald Schön, was the young Arthur D. Little consultant discussed in chapter 2. At the time, Schön was a thirty-three-year-old itinerant on his fifth career. He had already been a jazz pianist, military officer, professor of Deweyan philosophy, product designer, and consultant. At Commerce, Schön led innovation's new organizational home, the recently created Institute for Applied Technology (IAT). Located on the NBS's new campus in Gaithersburg, Maryland, the seven hundred employee IAT was to "stimulate the application of science and technology to national needs" by collecting and disseminating knowledge across boundaries between and within government and industry.[13]

Hollomon's second innovation expert, Daniel De Simone, also thirty-three, was a former Bell Laboratories patent lawyer. The son of Italian immigrants who had fled fascism, De Simone earned an electrical engineering degree from the University of Illinois, a law degree from New York University, and military stripes in the Air Force during the Korean War. Unlike Schön, De Simone would serve as a behind-the-scenes innovation expert for three decades. De Simone was put in charge of a new Office of Invention and Innovation within the IAT. The Office revitalized existing programs such as the NIC, a group of independent inventors and entrepreneurs created as a "brain bank" during World War II. De Simone turned what had become an honorary club into a consulting group of leading entrepreneurs and policy experts such as Polaroid founder Edwin Land and young RAND Corporation analyst Richard R. Nelson. His office also developed new initiatives such as the State Invention Program, which connected inventors and manufacturers.[14]

The bureaucratic innovators' first attempt to "put science to practical use" was an ambitious but failed effort called the "Civilian Industrial Technology Program" (CITP). The program had two parts: first, a university extension service for high technology that would redress the unequal distribution of federal research dollars which at the time went to a small group of connected states and institutions; and second, a federal assistance program to modernize traditional industries including construction, textiles, and urban transportation. But CITP quickly ran into opposition, both from the industries that the program hoped to aid and from a spat with New York Congressman John J. Rooney, who despised Hollomon's technocratic arrogance.[15]

Undeterred, Hollomon saw innovation as a solution to poverty and infrastructure modernization. After Kennedy's assassination, Hollomon integrated innovation initiatives into President Lyndon Johnson's Great Society, working with a coalition of liberal and centrist congressmen to pass legislation enabling his signature initiative, the State Technical Services (STS).[16] STS brokered public–private partnerships across all fifty states based on the nineteenth-century land-grant model.[17] Projects supported by STS included short courses in engineering management, information services on sewage waste treatment, and matchmaking between small businesses and university experts. President Johnson declared that the bill would "do for American businessmen what the great Agricultural Extension Service has done for the American farmer." In the process, the Act would "prevent more Appalachias" in which entire regions of the country were left behind in an innovation age.[18]

Hollomon also applied his vision of innovation to the government itself. He created internal programs such as the Science and Technology Fellowship to build a cadre of scientific experts who knew their economic ABCs. Johnson's aide Joseph Califano, Jr. described the program as part of a larger fusion of "the politics of innovation and the revolution in government management it has inspired."[19]

Commerce's innovation initiatives were building blocks of what the Johnson administration came to call "Creative Federalism." They aimed to create an interdisciplinary approach that linked the federal government to states, municipalities, and private industry, a model employed in Great Society programs such as Head Start and Model Cities.[20] However, the most lasting work of Hollomon's innovation offices was conceptual. He and his staff generated new knowledge about the innovation process to "establish a national climate that nurtures and cultivates technological creativity."[21]

INVASION AND UNREASONABLE MEN

Hollomon and his team drafted an enduring blueprint for American innovation. They emphasized the insurgent nature of innovation, the human qualities of innovators, and the appropriate policy environment to harness both for national progress.

The first guiding theme of the bureaucratic innovators was "innovation by invasion." According to Schön, the greatest source of change in existing industries came from external challengers. He first outlined the concept in a report that had doubled as his federal job application. In 1962, Hollomon contracted Arthur D. Little to survey the impact of "technical innovation" on American industry.[22] Foreshadowing the idea of disruption made famous decades later by Clayton Christensen, Schön's anonymously penned report synthesized hundreds of past client files to argue that innovation was a source of economic and social uncertainty. The government, however, could harness innovation's benefits for the public good and mitigate the disruptive effects of technological change.

Schön looked at three "lagging" industries ripe for modernization: textiles, construction, and appliances. Each was either stagnant or had been in economic decline for decades; that is, until they were threatened by new players from adjacent industries. Schön contrasted the dynamics of these traditional industries with the rapidly evolving semiconductor industry, which appeared to defy historical patterns with a new trend of continuous and accelerating change. His findings provided the framework for the government's STS program and the basis for future investigations of innovation ecosystems.

The bureaucratic innovators' second major focus was the cultivation of new generations of innovators. De Simone was especially concerned with the character traits of independent inventors. In Congressional testimony on the patent system, he expounded on the contributions that these "unreasonable men" had made to American progress.[23] The "inventor-innovator-entrepreneur," such as Polaroid's Land, was the rarest of this breed. Such men could bring ideas from conception to use. Other variations of innovator were independent inventors who coordinated with institutions, such as Samuel Ruben, inventor of the dry-cell battery, and corporate intrapreneurs in research and development labs. De Simone compared inventors and innovators to artists and writers such as William Faulkner and Ernest Hemingway. Both groups had a disdain for formal education and persisted in their goals where others gave up. Innovators were driven by the promise of financial reward; but like artists, they had a psychic need for recognition by peers and the public.

The Department of Commerce's vision of an innovation nation came together in a 1967 report that it distributed to every US congressman and state governor. The document *Technological Innovation: Its Environment and Management* came to be known as the "Charpie Report" after its chairman, Union Carbide president Robert A. Charpie. However, it was largely conceived and written by De Simone, who assembled a group of industry leaders in science, engineering, patent law, and business management.

The Charpie Report asked what the federal government could do to enhance innovative activity in the United States. It started with a broad definition of innovation as "the totality of processes by which new ideas are conceived, nurtured, developed and finally introduced." Innovation happened everywhere, from product development to society's ability to "adapt itself to the world or the world to itself."[24]

Despite the inclusive framing, De Simone's committee constrained its scope to market-driven technological change. The report's cover juxtaposed an illustration of an abacus and a digital computer to convey the world's ongoing knowledge explosion. It highlighted the accelerating adoption rates of new technologies such as television, jet travel, and computing. And it championed the dramatic economic growth of "technologically innovative companies" compared to the national average (16.8 percent to 2.5 percent). Within this narrower frame of technological innovation, the Charpie Report delineated invention from innovation as the "difference between the verbs 'to conceive' and 'to use.'"[25]

Fostering innovation required understanding a complex ecosystem and the government's role within it. Innovation was a regional phenomenon. Its main ingredients included universities, venture capital, entrepreneurs, and lines of communication between them. Innovation, moreover, was self-reinforcing—successful entrepreneurs helped to spawn additional entrepreneurs.

De Simone's committee critiqued the linear model of innovation and its focus on basic research. Using "rule of thumb" surveys, it declared that less than 10 percent of the cost of the innovation process took place in R&D laboratories, despite comparatively large federal investment. The essential unit for investigating innovation was not the university or government laboratory, but rather the private company, particularly small businesses. The committee's heroes were small start-up firms created by entrepreneurs, which suffered the most from policies created to regulate large corporations.

Innovators were at the core of the nation's innovation system. "The people who power invention and innovation" followed the classic typology laid out by Everett Rogers's *Diffusion of Innovations*. They were technically knowledgeable, cosmopolitan, risk-taking individualists who would stop at nothing to see their ideas succeed. Expanding on De Simone's examples of unreasonable men, the report showcased the work of independent inventors and small organizations behind dozens of technologies from air conditioning to the zipper. And it explored the common challenges these innovators faced such as lack of capital, lack of business training, and a surplus of risk-averse naysayers.

The Charpie Report's core message was "an abundance of ignorance" about the nature of innovation. The committee lamented a lack of existing research, which it used to justify basing its conclusions on personal experience. The group argued that harnessing innovation for national progress did not require substantial policy changes; however, it demanded a change in *attitude* and *environment*. Innovation was a "foreign language" with rules, uses, and learned habits. The government therefore had an important role as the nation's literacy teacher. Bureaucratic innovators could help Americans "*learn, feel, understand* and *appreciate* how technological innovation is spawned, nurtured, financed, and managed into new technological businesses that grow, provide jobs, and satisfy people."[26]

TOWARD A NATIONAL INNOVATION SYSTEM

Revolt, rather than renewal, was the prevailing national mood when Hollomon's team released the Charpie Report. Its architects worked on the optimistic side of a widening societal chasm. From the Vietnam War to environmental pollution and civil rights, many commentators had come to identify science and technology as root causes of the nation's unrest.[27] At the same time, a surging conservative tide that culminated with the election of Richard Nixon eroded over a decade of liberalism.

Hollomon was a casualty of these shifting political fortunes. His deputy, Schön, had already left to start a social innovation nonprofit and later accepted a position at MIT's Sloan School. In the waning days of the Johnson administration, Hollomon also departed, fleeing to a new position as the president of the University of Oklahoma.[28] His key initiative, the STS, also unraveled. Initially authorized to receive over $60 million in funding between 1966 and 1968, STS received less than $10 million. Hollomon's departure stood as a warning to future reformers of the perils of being perceived as a bureaucratic innovator.

Innovation initiatives, however, did not get thrown out with the liberal bathwater. On the contrary, by 1972, the Nixon administration made innovation a major priority. Why? For one, De Simone stayed in the new administration and he and other career bureaucrats took on greater responsibilities under Nixon. Likewise, many of the industrial scientists, patent lawyers, and entrepreneurs Hollomon had tapped as advisors continued to serve. Even as Nixon's staff reinterpreted how technology ought to be applied to civilian needs, they extended Hollomon's ideas into new federal agencies.

During Nixon's first term, science and technology policy nonetheless was marked by uncertainty and instability. The scientific community viewed Nixon's victory as the end of a golden age. For over a decade, scientists had received billions of dollars under the premise that basic science would generate tangible social benefits. In the process, scientists found themselves in the inner circle of federal policymaking. But the Vietnam War galvanized opposition among scientists, who organized by the hundreds to campaign against Nixon. Scientists, moreover, increasingly were targets of conservative politicians. Much of the animosity was cultural—scientists were typically cosmopolitan intellectuals or European émigrés with liberal politics. Although Nixon did not in fact slash funding for scientific research, the scientific community saw itself as beleaguered from multiple directions.

Research administrators in engineering and applied sciences, in contrast, found opportunity in the Nixon era.[29] Some, like William Baker, the head of Bell Laboratories, had long advised Republican politicians. Others, including systems engineer Simon Ramo, emerged as spokesmen for taming technology's "unintended" effects. Meanwhile, Nixon's Domestic Council pushed for a "business" approach to science policy. In this vein, Nixon appointed maverick aerospace engineer and businessman William Magruder to lead the New Technology Opportunities Program, a scheme for growing the civilian R&D budget. But the so-called "Magruder exercise" never received financial support and became infamous for a botched attempt to develop an ecologically friendly lawnmower.[30]

The Nixon administration sought to clarify its disjointed science policy by elevating innovation as its organizing principle. Ideas and programs long circulating in Commerce began to be taken up by the NSF and the White House's Office of Science and Technology (OST). The administration's "New Federalism" approach bore much in common with the Creative Federalism of the Johnson administration. It emphasized technology transfer to the civilian economy over further space and defense research. It expanded on the Charpie Report's image of an entrepreneurial nation thwarted by complex structures and unintended consequences that no one fully understood. New programs were to remove that ignorance through research into the innovation process, investment in incentives, and the elimination of regulatory "barriers" to innovation. The key difference in Nixon's vision of innovation was a stronger emphasis on the role of markets in social change. Hollomon's goals of innovation services through land-grant universities remained, but the focus shifted toward the "start-up" problems of new companies and the elimination of regulatory barriers to new innovations.

Two senior bureaucrats guided innovation efforts inside the Nixon administration. The first, Edward E. David, was the nation's first science advisor to come from a career in industry. Like Hollomon, David had been a corporate research director (at Bell Laboratories). The second, H. Guyford Stever, became the NSF's new director in 1972. Stever, the first president of Carnegie Mellon University, was a longtime Nixon supporter who fashioned himself as a scientific everyman, more at home hunting and fishing than in the cosmopolitan circles of his academic peers.

The Department of Commerce led the first of two parallel innovation initiatives launched by Nixon—the Experimental Technology Incentives Program (ETIP). Its director, Lewis Branscomb, was a young physicist in the NBS who spoke boldly for enhanced engagement between science and society through evidence-based industrial partnerships, deregulation, and new market creation. ETIP helped Branscomb launch a lifelong career as an innovation expert. He designed the initiative as an "innovative blend of 'science' and 'politics'" that would act as a "collaborative problem-solving" agent for improving government–industry relationships. ETIP's mission included programs for helping bureaucrats become change agents; research on the impact of deregulation in the railroad, pesticide, and pharmaceutical industries; and studies of the nature of innovators.[31]

The NSF led the second initiative, the Experimental R&D Incentives Program, designed to help bring science to market. Its director, C. B. Smith, was one of the few NSF employees with industry experience. He assembled

a group of social scientists to examine the efficacy of public–private partnerships, the impacts of tax incentives, and the value of prizes to stimulate innovation. Still, the NSF was cautious about making overtly political
choices. The agency described Experimental R&D Incentives as a source
of nonpartisan, objective advice that could assess empirically how innovation worked. In other words, Experimental R&D Incentives was intended to
make innovation scientific through assessment and accountability.

Nixon announced this suite of innovation programs during his 1972
reelection campaign. In the nation's first Special Message to Congress on Science and Technology, the president praised science but argued that "the mere
act of scientific discovery alone is not enough." Future prosperity required
combining "the genius of invention with the skills of entrepreneurship, management, marketing and finance." Instead of continuing to invest in basic science and classified weapons research, Nixon's plan would target the domestic
economy to bring "together the Federal Government, private enterprise, State
and local governments, and our universities and research centers in a coordinated, cooperative effort to serve the national interest."[32] His administration
would authorize the NSF to work on research applied to national needs, it
would reform patent policies to make federal discoveries available to private
firms through licensing, and it would initiate a series of incentives to enhance
the climate for innovation. Finally, the president himself would honor a small
group of creative Americans with an award on par with the Nobel Prize.

A NOBEL PRIZE FOR INNOVATION

Governments have long used prizes to strengthen ties between science,
commerce, and the nation. These incentives typically are one of two kinds:
challenge prizes and honorific awards. Challenge prizes, sometimes referred
to as inducement prizes, target major unsolved problems to stimulate private investment and widen the pool of possible solutions. Such awards
date to the famed Longitude Prize of 1714, in which the British Parlia-
ment offered 20,000 pounds for a practical means of measuring longitude
to within half a degree.[33] Honorific prizes that reward past achievements
also emerged with state-sponsored science in the eighteenth century. These
prizes seek not only to spur scientific growth, but also to elevate national or
imperial heroes of progress.[34] Honorific prizes gained newfound support in
the United States after World War II, with the establishment of the National

Medal of Science. First granted in 1962, the Medal represented science's—and the scientist's—ascendency in American society.[35]

The bureaucratic innovators who worked with Nixon's inner circle to craft the Presidential Prize for Innovation had briefly considered a challenge prize. They discussed a $100 million award for coal gasification as an energy-reducing innovation that nonetheless preserved the dominance of the fossil fuel industry, but they quickly abandoned the idea. Elevating innovators as American heroes, they concluded, was a more achievable goal.

The prize team navigated ambiguous beliefs about innovation's nature. The group asked: Were research and development breakthroughs more or less innovative than the implementation of social programs? Did innovation more typically come from large organizations or from creative individuals? What, for that matter, constituted a *national* innovation? Did a prize-winning innovation need to result from government investment?

The prize served to define what counted as innovation and who was an innovator during a transformational moment in United States science policy. For Nixon's Domestic Council, the prize would honor social or institutional returns on public investment. For Stever and the NSF, the prize was about retaining and reframing their identity as the country's source of scientific progress.[36] For Branscomb's new ETIP program, the prize would "encourage the best young scientists" to work in applied research.[37] Lastly, for De Simone, the prize was a culmination of Commerce's innovation advocacy, offering exemplars of American innovators while also building cross-cutting ties across government and industry.

What counted as an innovation? Winning submissions needed to address a significant national challenge and show wide public benefit, their innovations had to be in actual use, they had to be pioneered within the prior fifteen years, and they had to result from or be "verified" by scientific research. Up to ten prizes would be awarded in categories emphasizing the societal benefits of innovations:

1. Environmental quality
2. Energy
3. Natural resources
4. Health care and safety
5. Food and nutrition
6. Education
7. Housing and community development
8. Transportation
9. Communications and information processing
10. Productivity and international trade[38]

Eligible innovators included both people and organizations.[39] However, the prize committee placed a strong emphasis on individuals and start-ups. David, in particular, pushed the optics of the *"little guy,"* through whose "inspiration . . . others may say to themselves 'By golly, maybe I can do it, too!'"[40]

The innovation prize was also an attempt to redefine the image of science. In contrast to the Medal of Science, an honor for lifetime achievement with no financial reward (which Nixon conspicuously failed to award in 1971), the innovation prize would come with a cash award as a "stimulus to creative endeavor." In this aspect, the prize also reflected Nixon's disdain for elites. The competition would be open to anyone, rather than only those nominated by scientists. While David was careful about balancing the innovation prize with the Medal of Science, Nixon's Domestic Council viewed the Medal of Science as a reward for the president's liberal opponents. Political staffer Ray Waldmann asserted that "our intention *should be* to compete with and outstrip other prizes."[41]

De Simone assigned a team of scientific bureaucrats to carry out the competition. George Arnstein, a forty-year-old nuclear physicist, managed the nuts and bolts. He was joined by counterparts in an alphabet soup of agencies with a stake in innovation—OST, NSF, Commerce, AEC, NASA, the Department of Health, Education, and Welfare (HEW), and the newly formed Environmental Protection Agency (EPA).[42]

The prize's rollout was rushed and disjointed. Nixon publicly announced the prize before deciding how the awards would be funded or what would count as a national innovation. Science and engineering societies received requests for nominations less than two weeks prior to the prize deadline. The competition also was not advertised in national media nor to groups such as the NAACP, whose advocacy might have broadened the pool of potential women and Black innovators.[43]

Despite problems, nominations for the innovation prize poured in. After a short extension, the competition yielded nearly five hundred entries that ranged from scientists and engineers to corporations, university departments, garage tinkerers, and Catholic nuns. Innovations included antibiotic drugs, chemical processes, solar heating devices, holography, numerous advances in computing, the Saturn rocket, the Bay Area Rapid Transit System, environmental monitoring systems, packaging plastic, and a program for rehabilitating convicts (led by the nuns).

In the first round, the interagency bureaucrats whittled the nominees down to forty-three cases. Many well-known scientists, engineers, and designers failed to advance, including Buckminster Fuller (geodesic domes), Barry Commoner (ecology), and Paul Ehrlich (population growth). Those surviving the first cut included Norman Borlaug (high-yield wheat), Jack Kilby (integrated circuits), and Charles Townes (masers and lasers).

In the second round, external advisors identified fifteen finalists. The external advisors primarily were PhD-trained industrial scientists from Bell Laboratories, Monsanto, and the like. Many, such as Charpie, had served in federal advisory roles throughout the 1960s.[44] In its deliberations, the group determined whether one's innovation had achieved widespread use, and called on the Patent Office to examine priority claims.[45] Unsurprisingly, every one of its choices was a scientist or an engineer. All were white men. The industry advisors rejected a potential sixteenth case, Sesame Street, by a vote of five to four.

In August 1972, the White House selected six winners: Samuel Ruben, an independent inventor in the Edisonian tradition; Harold Rosen, an aerospace pioneer; Willem Kolff, a biomedical engineer; Edward F. Knipling, an entomologist and environmentalist; John Backus, a computer scientist; and, overturning the advice of their advisory board, Joan Ganz Cooney and Lloyd Morrisett, creators of *Sesame Street*.[46]

LIVES OF THE INNOVATORS

None of the intended recipients of the Presidential Prize for Innovation are household names today. Even at the time, they were unfamiliar to most Americans. The prize was designed to rectify this; by elevating their achievements it promoted a uniquely American social type and showcased the government's central role in powering innovation.

By reconstructing the life stories of innovators from the committee's case files, we can see how a national innovation system was to be manifest in flesh-and-blood morality tales. The prize committee intended to distribute a series of popular biographies for each winner that reinforced the human qualities of innovation. Selected winners encapsulated the political and personal visions of the bureaucratic innovators. Individually, each winner captured a distinct persona of the government's lessons for incentivizing innovation. Collectively, these innovators' life stories told of a dynamic nation that balanced expertise with democracy, infrastructure with novelty, and private enterprise with public good.

Here are those lessons as the Nixon administration intended them.

THE VANISHING INDEPENDENT: SAMUEL RUBEN

The prize's eldest recipient, Samuel Ruben, exemplified nostalgia for the innovator as an unreasonable man. Born in 1900, Ruben never attended college. He joined his first start-up, a munitions company, at the age of 17. When the business failed, Ruben established a personal R&D laboratory. Instead of creating his own company, Ruben licensed his patents (over three hundred in his career) to P. R. Mallory and Company.

Ruben's greatest achievement was the miniaturized dry-cell battery. During World War II, durable, compact, and long-lasting batteries were in demand for walkie-talkies. Their use, however, was limited by a shortage of manganese ore and by the challenges of operating under the environmental extremes of combat. Working with the National Inventors Council, Ruben created a mercury-based dry-cell battery that resolved both challenges. It was just half an inch long, half an inch in diameter, and a quarter ounce in weight. Perfected over decades, Ruben spawned a multibillion-dollar industry with applications ranging from cardiac pacemakers to the electronic wristwatch. On the strength of Ruben's batteries, P. R. Mallory rebranded its product as Duracell in 1965.

In an R&D environment dominated by PhD-trained scientists in corporate or federal laboratories, Ruben stood for what historian Eric Hintz calls the "vanishing American genius."[47] He was an advocate and symbol of the patent system and the need for its reform to better serve entrepreneurs.[48] Although Ruben criticized the anonymity of independents like himself, between 1964 and 1970, he racked up honorary doctorates and lifetime achievement awards. According to Ruben's award citation, the inventor was living proof that "the age of the creative, independent" was not over.[49]

THE SPACE AGE GENIUS: HAROLD ROSEN

Where Ruben harkened to an Edisonian past, communications engineer Harold Rosen exemplified the Cold War's entrepreneurial possibilities for civilian technology. His life story celebrated the success story of satellite communications, but also provided a lesson on the need for government reform of the military–industrial complex and corporate support of internal innovators.

Born in 1926, Rosen was an electronics wunderkind who graduated from high school at fifteen. Like most industrial scientists of his generation, Rosen fell into defense research after World War II. He earned his PhD in electrical engineering at Caltech while working part time at the aerospace firm Raytheon on guided missiles and radar.

The launch of Sputnik changed Rosen's life. The newly minted PhD was developing antiaircraft radar for Hughes Aircraft. However, the intercontinental ballistic missile that carried Sputnik threatened to render his work obsolete. He and colleagues Thomas Hudspeth and Don Williams turned to space.

Rosen and his team designed a satellite that maintained a fixed spatial position, allowing it to serve as a global communications hub. Their solar-powered prototype weighed just fifty-five pounds. By July 1963, Rosen's satellite was in orbit. That August, President Kennedy conducted the first live satellite call with a foreign head of state. The following year, a second satellite broadcast the Tokyo Olympics. Over the next thirty years, Rosen aided in the design and launch of over 150 civilian communications satellites.

Rosen's innovation was a case study of research applied to national needs. Satellites were a "dual use" technology that applied defense expertise to the civilian economy. And yet Rosen's biography was also an indictment of the military–industrial complex's bureaucracy. Hughes management doubted that the company could sell more than an hour a week of satellite television and only after Rosen threatened to quit did the company embrace the project.[50] His prototype, however, challenged a project by a DOD- and NASA-funded consortium involving multiple defense contractors. In contrast to Rosen's elegant and inexpensive prototype, the consortium's satellite weighed 1,200 pounds, required a custom-designed rocket, and was plagued by cost overruns. Rosen received funding only because he had a well-placed champion at DOD.

For the selection committee, Rosen offered multiple insights on the path toward a more innovative nation. First, there were many driven, risk-taking individuals working in large organizations. Second, the conversion of military to civilian research could produce significant commercial benefits. Finally, the government needed policies for streamlining and new ideas throughout the federal bureaucracy.

THE IMMIGRANT BIOENGINEER: WILLEM KOLFF

Willem Kolff, the father of artificial organs, demonstrated how entrepreneurial frugality combined with creative thinking saved lives. His award was also a parable of how the United States was in a global contest for talent and a statement about how the benefits of medical innovation should be distributed in a market economy.

The Danish-born doctor began work on his prize-winning artificial kidney in 1939. He conducted his first experiments with cellophane tubing from a sausage casing and an automobile water pump. He continued to improve his process during World War II while resisting Nazi occupation.

In 1945, Kolff's artificial kidney saved its first patient. Kolff then distributed drawings and prototypes of his machines to hospitals in the United States and elsewhere. In 1950, he immigrated to the United States. At the prestigious Cleveland Clinic, he continued to improve his dialysis technology. With funding from a real estate financier, he partnered with the Peter Bent Brigham Hospital to develop the Kolff Brigham Artificial Kidney. He then invented a disposable kidney that decreased the cost and increased the accessibility of his life-saving treatment.

Kolff's career spanned dramatic changes in biomedical research. Bioengineering was just emerging as a discipline with new institutes and departments. In 1967, Kolff joined the University of Utah's new Institute for Biomedical Engineering. The culture and financing of medical research was also shifting toward patenting and corporate partnerships. Here, the selection of Kolff presented a negative lesson. The pioneering doctor later lamented that he did not pursue patents on his various dialysis machines; at the time, however, to do so was considered unethical for a member of the American Medical Association.

For the Presidential Prize committee, Kolff's selection finally highlighted a countervailing dimension of the government's role in medical innovation: Who pays for it? Despite strides made by Kolff and others, treatment for kidney disease remained an expensive, technology-dependent procedure with significant disparities in treatment. One week prior to the 1972 election, after years of patient advocacy and political divides over healthcare reform, Nixon made dialysis the first federal entitlement ever to cover a specific medical diagnosis.[51]

THE COWBOY ENVIRONMENTALIST: EDWARD F. KNIPLING

Edward Fred Knipling's prize selection elevated an overlooked American innovator—the land-grant scientist. This rare breed of stoic innovator eschewed politics, ideology, and even profit in the nation's service.

Knipling could have stepped off the page of an Everett Rogers case study. The son of an immigrant farmer, he grew up on a small farm on Texas's Gulf Coast. The humid, swampy region was plagued by the screwworm fly, which ravaged his father's cattle.

Fifty years later, Knipling would vanquish his childhood nemesis. Screwworm flies are subtropical insects that lay eggs in the exposed wounds of warm-blooded animals. Their flesh-eating maggots caused suffering, death, and tens of millions of dollars in agricultural losses annually. Attempts to battle screwworms faced two barriers: insecticide solutions were poisonous to livestock meant for human consumption and the use of insecticides could not keep up with the reproduction rate of the flies.

In 1931, after graduating from Texas A&M, Knipling took a job at the US Department of Agriculture. As he rose through the agency's ranks, he earned a PhD from Iowa State, oversaw the wartime development of DDT, became a skeptic of chemical insecticides, and pursued environmentally sound pest control.

In 1937, Knipling and his colleague R. C. Bushland studied the mating patterns of screwworms, recognizing that while females mate only once in their lifetimes, males mated many times over. Knipling pursued the sterilized male technique, seeking a pesticide-free solution to insect control that utilized the mating cycle of flies to gradually eradicate their populations. He developed population dynamics models, snuck into hospitals to experiment with X-ray machines, raised billions of flies in captivity, created an assembly line that sterilized them, and modified small aircraft to deliver them.

In 1954, at the behest of the Dutch government, Knipling eliminated a screwworm invasion on the island of Curacao. Livestock producers in Florida then convinced the USDA to attempt a similar program throughout the state. In 1959, Texas Senator Lyndon B. Johnson and Governor Price Daniel, both ranchers, organized a consortium of Southwestern states to fund a larger eradication effort. By 1966, Knipling eliminated screwworm flies from the United States.[52]

From Kennedy to Nixon, the federal government championed Knipling as a land-grant hero. USDA publicity described him as "more like a cowboy than the popular conception of an egghead scientist."[53] He earned the praise of environmentalist Rachel Carson and agribusiness executives alike. President Johnson awarded him the National Medal of Science. And, on his retirement in 1971, the Nixon administration honored him with the Distinguished Federal Civilian Service award. Despite these accolades, the innovation prize report emphasized a common refrain: because of the novelty of his ideas, he "received little support or encouragement" and had difficulty securing funding. "Yet, Dr. Knipling persevered."[54]

THE PEOPLE'S PROGRAMMER: JOHN BACKUS

By selecting mathematician John Backus as a national innovator, the Nixon administration sought to praise the role of creative misfits in a nascent digital revolution. Between 1954 and 1957, Backus led a team at IBM that developed FORTRAN (FORmula TRANslating System), the world's first high-level programming language. Nearly every computer manufacturer adopted their innovation, which redefined the labor practices of computing and contributed to innumerable scientific discoveries.

According to Backus, information technology's "mother tongue" sprung from laziness.[55] The son of a wealthy stockbroker, Backus dropped out of the University of Virginia, reportedly out of boredom. He was subsequently conscripted to fight in World War II. He later tried medical school but lasted just nine months. He finally found his calling in a math graduate program at Columbia. As Backus approached graduation, he was enticed by a computer display in the window of IBM's Madison Avenue headquarters. He walked in, mentioned his mathematics training, and was whisked upstairs, challenged with logic puzzles, and offered a job.

FORTRAN was an infrastructural innovation that solved one of computer science's most difficult problems. Programming was an arduous task done in machine language by an army of specialized workers like Backus. Programming—and the associated work of debugging—accounted for up to three quarters of the cost of owning a computer. In the early 1950s, "automatic" programming in which the computer translated natural language into machine language seemed an impossible challenge, with early compilers slowing processing power by a factor of ten.

Backus, who spent his days on defense research calculations, imagined a better way. He convinced his manager to explore the idea of a machine translator for the company's new 704 model computer. As the work progressed, his Program Research Group grew into a diverse team of proto-computer scientists, most in their twenties and early thirties. The team defined their language's commands, developed a compiler to translate them, spoke to disbelieving customers, spent hundreds of hours debugging, and developed a *Programmer's Primer* for users. In April 1957, IBM released the first version of their new language. By 1972, FORTRAN was the gold standard of software, a term that had not existed at the project's start.

Information technology represented the most visible domain of technological innovation at the time of the Presidential Prize and the prize committee used Backus's award to highlight how fundamental questions of computer science were closely linked to market needs. Backus's initial pitch for the project was what today's innovation experts call a *value proposition*: "Can a machine translate a sufficiently rich mathematical language into a sufficiently economical program at a sufficiently low cost to make the whole affair feasible?"[56] Backus's answer applied research to national needs. FORTRAN was both a management innovation and a scientific "enabling" tool that made computing accessible.

NOT LIKE THE OTHERS: JOAN GANZ COONEY AND LLOYD MORRISETT

Sesame Street was more than a children's television program in 1972. Depending on one's vantage, it was an educational "head start" for every child or an abandonment of faith in public schools; a scientifically crafted experiment or a market-driven entertainment; "a historic step" in the use of mass media to improve humanity, a corporatist sell-out, or a socialist scheme. The prize committee was not sure what to make of it.

Sesame Street started as a napkin sketch. In 1966, at a dinner party among philanthropists and Manhattan television producers, Joan Ganz Cooney and Lloyd Morrisett came to talk about television's potential as a teaching tool. Cooney had spent more than a decade working for public and commercial networks. Morrisett, trained as a psychologist, had a longstanding interest in early childhood education; and, as an employee of the philanthropic Carnegie Corporation, he led a nationwide student assessment initiative.

Cooney and Morrisett's vision for *Sesame Street* tackled two societal problems simultaneously: first, race and class-based educational disparities, and second, the negative impacts of commercial television on childhood development. The pair of staunch liberals were frustrated by the limited gains of Great Society programs. Cooney had long been interested in early childhood education; in 1964, she produced a federally funded documentary about innovative preschools that were precursors to Johnson's Head Start program.

She wanted Head Start initiatives to reach all children but came to doubt the government's practical skill and political will. At the same time, by 1966, television was ubiquitous; Ninety-six percent of American households had televisions and it was estimated that the device was on sixty hours a week in the homes of the working class and urban poor.[57] Yet the private advertising model that supported the medium fostered commercialism and violence. Cooney and Morrisett saw an opportunity to address both challenges by blending educational research with techniques of corporate entertainment.

Cooney and Morrisett were as much entrepreneurs as educators. Cooney had first approached New York's public television station WNDT with her idea for an educational children's program, but the network rejected the idea. Along with Morrisett, she convinced the head of the US Office of Education to cover half the cost. The idea was for a novel public–private partnership with a commercial network. But no commercial broadcaster would pay. Instead, Cooney and Morrisett raised $8 million from government programs and private foundations. In the process, they focused on the prekindergarten child from "an urban disadvantaged community" as their primary audience.[58]

Cooney and Morrisett then formed the Children's Television Workshop (CTW), a multidisciplinary team that included Harvard educational psychologist Gerald Lesser; David Connell, the executive producer of the children's program *Captain Kangaroo*; and avant-garde puppeteer Jim Henson. Despite wildly different backgrounds, all were driven by the desire "to do something useful."[59] In the parlance of today's innovation experts, the CTW was a high-performing team. It was guided by a set of principles that valued collaboration, an age-appropriate curriculum, and rigorous assessment of the program's impacts.[60]

When *Sesame Street* premiered on November 10, 1969, it was an instant sensation. It inspired copycat programs on commercial networks and attracted criticism from many quarters.[61] The show also became a political liability. In 1971, Head Start director Edward Zigler worried that the Nixon administration would use the show as fodder to eliminate funding for his initiatives. He criticized the "tokenistic approach" to learning and withheld funds earmarked for the show.[62] At the same time, the show was the cause célèbre of the Action for Children's Television, a consumer movement that advocated for educational children's programming.

Sesame Street was the only Presidential Prize winner in which the White House overturned its selection committee. The show was one of two stellar educational candidates. The other was MIT physics professor Jerrold Zacharias, who had modernized the science curriculum of the nation's schools.[63] A scientist's scientist, Zacharias had been nominated for the prize by both David and Stever. Still, Big Bird really did have a friend in Nixon. Even as the president considered defunding PBS, First Lady Patricia Nixon was a major *Sesame Street* supporter. In 1970, she hosted the show's Christmas special from the White House. Zacharias, in contrast, had begun to organize scientists for Nixon's democratic challenger George McGovern.[64]

A worthy competitor does not explain why *Sesame Street* initially missed the national innovation cut. *Sesame Street* challenged the prize committee's understanding of innovation as inherently technological. While the show's approach to education rested on a scientific foundation, it was far better known for its portrait of empathy and shared humanity to repair racial and economic divisions. Cooney, moreover, was the only female nominee selected from a pool that included just thirteen women and she lacked a PhD in science and technology.[65]

What did *Sesame Street* exemplify about emerging innovation ideals? The show was collaborative, cross-disciplinary, children-oriented, and pioneering in its budget model. It also represented a new and controversial role for the government as a creator of entertainment. It deliberately used technological media as a tool of social change. It was simultaneously communal and corporatist. Of all the would-be Presidential Prize winners, the show was closest to the theory of innovation outlined by Schön in 1963. *Sesame Street* was a transgressive innovation by invasion.

BIPARTISAN MYTHS

On September 15, 1972, instead of celebrating the nation's innovators, the White House received the first indictments in the Watergate scandal that brought down Nixon's presidency. The prize ceremony was postponed. Then postponed again. With the exception of an inquiry in *Science* asking "whatever are the presidential prizes?" the award disappeared with little notice.[66]

Watergate was one of many factors in the prize's demise. Uncertainty about which agency would pay for the awards and whether a deadlocked Congress would approve the federal budget in time to grant them also

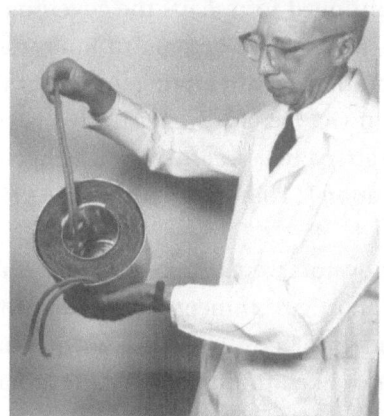

FIGURE 3.2

Selected winners of the Presidential Prize for Innovation. Top: Samuel Ruben and Harold Rosen. Middle: Willem Kolff and Edward Knipling (center). Bottom: John Backus and Joan Ganz Cooney and Lloyd Morrisett (Samuel Ruben orphan rights; Harold Rosen courtesy of Boeing; Willem Kolff courtesy of Spencer S. Eccles Health Sciences Library, University of Utah; Edward Knipling by USDA National Agricultural Library; John Backus courtesy of IBM; Joan Ganz Cooney by Lynn Gilbert; Lloyd Morrisett by Joi Ito).

played a part. This prompted Stever to keep the competition quiet because he worried that the prize would raise red flags in Congress and jeopardize the NSF's Experimental R&D Incentives Program.[67] Nixon's inner circle, moreover, was so anxious about negative press that they contacted winners to see if they would accept their awards.

The litany of missteps signaled disarray in the government's nascent innovation policy. In January 1973, David resigned as OST head, lamenting that his ideas were not heeded. Rather than replace him, Nixon abolished the entire office. Stever then assumed a dual role as White House science advisor and NSF director, where he continued to reshape the government as an innovation catalyst. De Simone, for his part, became an innovation czar under Stever and earned minor notoriety for attempting to convert the United States to the metric system.[68] In October 1973, a distracted Nixon granted the National Medal of Science after a two-year hiatus with a speech that failed to mention the word innovation.[69] Four months later, the president resigned.

The setbacks of the bureaucratic innovators were temporary. Some, such as Hollomon, Schön, and David, had short government tenures; but they continued to advocate for innovation from positions in academia and think tanks and courted the attention of likeminded counterparts around the world.[70] Others, like De Simone and Branscomb, spent their careers in government and ascended to prominent bureaucratic positions. They viewed their work as rational, apolitical, underappreciated, and incomplete as they shaped federal innovation ideals through the ensuing decades of partisan swings.[71]

But what of the innovators who never had their moment on the front stage? Personifying innovation was an important part of the bureaucratic innovators' strategy for social and political change. To the bureaucratic innovators, these human agents revealed vital patterns in the cultural relationships that made innovation possible—the interplay between individuals, companies, scientific discovery, economic growth, social adoption, consequences, and appropriate ways of living. They needed to be identified, cultivated, and rewarded.

The winners selected by the Nixon administration represented an embryonic vision of innovators as national heroes. Nixon's innovators, however, were not slick jetsetters. They were entrepreneurial, but not necessarily entrepreneurs. They worked within government or in partnership with it. Many of their innovations were infrastructural and had succeeded only with government investment. They presented a society in which creative individuals, corporations, and government agencies all had a part in making a better world.

The prize caper, however, revealed the tenuousness of innovation's myth of civic harmony. Not even historians are likely to recognize most of Nixon's would-be winners.[72] Instead of the eradicator of screwworms or the anti-Nazi tinkerer, today's heroes of innovation are the digital disruptor and the billionaire entrepreneur. Amid the collapse of faith in government, a more robust vision of innovation was taking shape along Boston's Route 128, in Silicon Valley, and on the pages of New York magazines.

4 BE AN INNOVATION MILLIONAIRE!

It is less and less possible to explain who I am in terms of the job I do, the profession I represent, the region of the country in which I live, the institution to which I belong or the class or race from which I come. This newly experienced difficulty in saying who I am goes hand in hand with an increasing confusion about where I am going.

—Donald Schön, 1967

This is the most incredible thing to be—not an astronaut, not a football player—but this.

—Steve Jobs, 1985

In January 1970, two hundred technology managers met at a secluded mansion in Glen Cove, Long Island. Their mission: to learn what it takes to be an innovator. From the comfort of their rooms, executives from AT&T, Honeywell, IBM, and 3M talked shop via closed-circuit television and telephone with leading entrepreneurs, science administrators, and academics. In this proto-Zoom meeting, the gurus paced the stage of an intimate theater as they wove parables about how their lives had been changed by the "accelerating rush of innovation."[1]

The pricey workshop was the brainchild of a new media start-up called "Technology Communication, Inc." The weekend event captured the club-like exclusivity, expert insight, and collective self-help for revolutionary times that the new venture aimed to cultivate. Each evening, the speakers again held court in the bar, offering access into a brotherhood called "The

Innovation Group." Recruitment advertisements in the *Wall Street Journal* and *Scientific American* declared that those who joined would be "a set of men very important to this country."

Acceleration was on these men's minds. Technology managers were being forced to reinvent themselves and their companies for an unsettling age. In the previous two years, they had witnessed the Apollo moon landing, anti-Vietnam War protests, sweeping environmental regulations, landmark civil rights legislation, political assassinations, rising extremism, the exponential growth of the microelectronics industry, and the threat of international competition.

Technology Communication was selling a foothold for survival in a world of chaotic flux. In personal gatherings and in the pages of its monthly magazine *Innovation,* Innovation Group members would confront the promises and perils of technology while turning a profit and enjoying meaningful and creative lives. Together, they would cultivate a new art without "teachers or precedents or tradition."[2]

Today, this trope of the switched-on innovator is a cliché. It is one that historians associate with the utopianism of late-1960s California where high technology and hippie culture developed in tandem and then flourished in the personal computer revolution of the 1980s.[3] But this ideal is more deeply connected to the Cold War than is generally understood. It is just as much a product of the team behind Technology Communication who created new venues in which innovation experts exchanged ideas, fashioned the mythology of a digital revolution, and then defined, packaged, and sold it as the growth of the microelectronics industry redistributed the nation's centers of cultural and economic power from the East Coast to the West Coast.[4]

This chapter explores how the innovator as a creative capitalist who thrives amid endless change became a recognized and desired persona. It demonstrates how the abstract and unsettled identities of innovators posited in the 1960s coalesced in a set of virtues centered in technoscientific entrepreneurship that nonetheless made claims to the public good. By studying Technology Communication and its successors, I show how innovation's champions used the aspirational ideal of the innovator to navigate the upheaval of their circumstances. A central feature of their valorization was the reimagination of management as a creative act that blended expertise with passion in an ever-changing world. They set as their goal the development of a new way of living.

SCIENCE'S IMAGE PROBLEM

Representations of who scientists and engineers are and how they should act are as old as science itself.[5] The British Royal Society, for example, was created as much to police who had the moral authority to investigate nature as it was to exchange members' discoveries. As the enterprise of science grew, so did the spokespeople for the character of its members. Victorian popularizers instructed the right ways of living among scientists and engineers. Samuel Smiles, the pioneering *Self-Help* advocate, for example, used popular "lives" to promote engineers as the progress-seeking souls of British imperialism.[6] Early science fiction writers, most notably Jules Verne, likewise accentuated the moral authority of scientists in society. Research managers in the early twentieth century described corporate scientists as "industrial explorers" and in the spirit of Max Weber, identified their "vocation" to truth and economic growth.[7] After World War II, scientists achieved newfound celebrity in American culture for their development of the atom bomb and other breakthrough technologies while simultaneously stressing their commitment to the purity of knowledge.

At the height of the Cold War, American science again faced an image problem that hinged upon the character of its practitioners. To compete with the Soviets, the nation needed talent. To recruit that talent, it needed to convince its citizens that the scientific life was a worthy pursuit. Unfortunately, when, in 1957, anthropologists Margret Mead and Rhoda Métraux asked American high school students to describe scientists, they discovered a positive view of science combined with a lack of interest in scientific careers. *Science* was for mankind's benefit, but the *scientist* was a white-coated man in a laboratory, bald, tired, and unfit to marry. What was lacking, they concluded, was a true portrayal of the "real, human rewards of science—on the way in which scientists today work in groups, share common problems." They suggested that educators and media produce "pictures of scientific activities of groups, working together, drawing in people of different nations, of both sexes and all ages, people who take delight in their work."[8]

It was not just the American public that was uncertain about scientific careers. Scientists and engineers also wondered what kind of life they had chosen. Technical fields experienced staggering growth in the Cold War era fueled by the arms and space race with the Soviet Union and the booming market for consumer technologies. By 1960, engineering was the leading occupation for white-collar males in the United States. There were over 870,000 engineers in America, a nearly four-fold increase from before the war. Approximately 75 percent were employed in just 1 percent of all firms and corporations with a workforce of ten thousand or more employed 35 percent of America's engineers.[9] The rate of growth was even higher in the top ranks of American science. Prior to World War II, one in three PhD trained physicists worked in industry; by 1957, the ratio had risen to half. Scientists and engineers were also more likely to be working together and to be doing so in a corporate environment.[10] At the same time, universities had begun to resemble corporations with dedicated research centers and practitioners moving freely between academia, government, and industry. The lines between *science* and *engineering*, never distinct in the first place, were increasingly blurred.

A booming market of academic and popular commentary fed the hopes and concerns of the scientific enterprise. In the aftermath of the atom bomb, popular media contributed to making scientists national personalities. However, social scientists such as Robert K. Merton portrayed a conflict between the "scientific ethos" and corporate bureaucracy.[11] This popular fascination with science and technology coincided with explosive growth in the publishing industry, with specialty magazines increasing in circulation by over 300 percent between 1943 and 1963.[12]

Scientists and policymakers wanted to channel the public's interest, but they lamented that existing popular representations debased the value of science. They thus sought to support the right kinds of scientific journalism and to have a hand in producing them.[13] The marquee effort was *Scientific American*, which underwent a major rebranding after World War II to shape the public's understanding of scientists as well as that of "the scientists themselves, the doctors and engineers, the executives and managers of industry and those engaged in the non-technical professions of teaching and the law."[14] This form of popularization produced a new generation of science writers, equipped with technical backgrounds and a rolodex of top scientists, administrators, and policymakers to provide content.

CHANGE FOR SALE

In 1960, an enterprising publishing executive at the firm Conover-Mast saw an untapped opportunity in the drumbeat for scientific communication. *Scientific American* tended to look backward, using history to connect modern science to the scientific revolution. Science popularizers also cast scientists as an ivory tower guild that pursued ideas for their own sake. According to William G. Maass, however, the "consumption of fundamental science by technology" had accelerated such that development from "idea" to "utilization" had been reduced from decades to weeks. In a knowledge-based economy in which "discovery can occur anywhere," the lines between disciplines and nations were breaking down. So too were the identities of scientific practitioners. "That we need a single word to mean both engineer and scientist, that the two form a single community," Maass wrote, "is obvious."[15] To interpret and communicate this community of technoscientists, his magazine, *International Science and Technology (IST)*, targeted the leading 10 percent of the world's scientists and engineers: "the responsible decision-making technical men."[16]

Maass, whom a reporter at the Technology Communication workshop later described as a "supercharged, crackling blur of a man," spared no expense in realizing his vision. He assembled an exceptional team of science journalists. Most notable was *IST* senior editor Robert Colborn, who had spent a decade as managing editor of *Business Week*. Colborn was a civil engineering graduate of Dartmouth turned writer, who had penned the 1958 nuclear disaster novel *The Future Like a Bride*.[17] Maass also hired the famed designer Will Burtin to establish a visual identity for the magazine.[18] *IST* conveyed technoscience as the driver of an international revolution that was modern, creative, aesthetic, and democratic. Each issue's cover featured an abstract painting representing a scientific theme commissioned by *IST*'s art director who trawled Manhattan's galleries in search of talent.[19] Contributors were introduced with candid portraits and informal biographies. Articles included sketches in the margins to convey creativity in action. Instead of a "letters to the editor" section, *IST* had a Communications Center, for readers to pose interdisciplinary problems and receive answers from other readers. *IST* projected itself as less encyclopedic and more relevant than *Scientific American* and as more revealing about the practices and personal qualities of scientists than journals such as *Science*.

IST aimed to provide "state-of-the-art" reports authored by leading experts while also giving readers an understanding of the driving forces of their age.

Conover-Mast was an unlikely home for this journalistic experiment. In business since 1927, it was one of the largest industrial trade publishers in the United States. The company was founded when its partners Harvey Conover and Bud Mast left editing jobs at McGraw-Hill's *Factory* to start the competing *Mill and Factory*. Other titles in the company's catalog included *Boating Industry* and *Volume Feeding Management*. In Sputnik's wake, however, Conover-Mast evolved with the industries it covered. In 1958, for example, *Aviation Maintenance & Operations* was rebranded *Space/Aeronautics*. *IST* also had an unusual business plan: Conover-Mast gave it away for free to the majority of its 100,000 readers. To cultivate an audience, the magazine bled cash, reporting a net loss of $1.5 million its first year. It established credibility, however, with an editorial board of international scientific statesmen. Advertising from corporations, industrial research laboratories, and government agencies followed. By the end of its second year, *IST* was in the black.

IST poured its energies into answering the call for positive images of technoscience by linking its ethos to the lives of its practitioners. According to *IST*, "that horrible word—IDENTITY" was critical to combat simplistic stereotypes of mad geniuses or staid organization men and to capture a "sharp turn" in the "style of our society" visible in the "life of the scientist or engineer."[20] Individual quirks were deployed toward generalizable arguments about the scientific life. Readers learned for instance that Nobel laureate Albert Szent-Györgi once intended to become a doctor in the Dutch East Indies and started his biochemical work using apples and bananas. *IST* packaged these lives into a best-selling book, *The Way of the Scientist*, which emphasized a "new community" that was transforming society.

IST offered a then-unique focus on the infrastructure of science and technology that for the first time in popular media drew attention to the "innovation process." In addition to carrying interviews with Nobel Prize winners, the magazine worked with research managers to describe and analyze what they did. In 1964, Donald Schön distilled the theory motivating his work inside the Department of Commerce in the article "Innovation by Invasion." Schön claimed that the greatest source of innovation in an industry or research field came from outsiders who disrupted the status quo. He followed up in 1966 with the article "Fear of Innovation" that explained why institutions and organizations resisted change even when the failure to act had negative consequences.[21]

Another 1964 article of this sort, "From Research to Technology," altered the trajectories of both its author, Jack Morton, and the *IST* staff. The Bell Labs manager explained how he had guided the transistor from laboratory to market and how the experience had opened the door to a general systems theory of innovation.[22] In 1966, he followed up with "The Microelectronics Dilemma," which charted the decentralized expansion of microelectronics companies and the incessant creative destruction it generated. To "survive and grow in a new era of technology," Morton wrote, required mastery of the "people-process of innovation."[23] Summarizing Morton's insights about the development of the transistor and the revolutionary pace of the microelectronics industry, *IST* concluded that without the Mortons of the world, the transistor would be just one more undeveloped discovery, so it was vital to cultivate experts who understood innovation's vagaries and could balance creative exploration with practical results.

In bringing insights about innovation to readers, *IST* conveyed self-awareness of having tapped into world-changing energies. Maass marveled that *IST* was "riding a wave, a wave of technical and social change" in which the entire globe was "under several kinds of pressure to become even more international than we are."[24] In 1964, *IST* used the marketing slogan "Change or Die!" to showcase how technology was remaking society. This new epoch, he explained, was a capitalist one. It was competitive and "unforgiving of mistakes," however, for the "enterprise which is in tune" with the patterns of change, it was possible to "accelerate very very fast" with great reward.[25]

By the second half of the 1960s, however, Maass's wave of technical and social change resembled a tsunami. The Cold War scientific ideals that motivated *IST*'s creation appeared increasingly quaint and dated amid growing pessimism and hostility toward technology. *IST*'s editorial staff watched changing attitudes toward technology with unease and documented how it altered scientists' and engineers' conceptions of self. In 1967, for example, *IST* published a segment with the environmental activist Barry Commoner titled "The Eroding Integrity of Science."[26] Then, in a 1968 issue, Colburn editorialized that race riots stemmed from the fact that there would always be people "who can't fit themselves into technology's world," but he contended that "the newest technology may actually be reintroducing a human scale into the mechanisms of society."[27]

Out-of-control technology was a societal concern, but it was also a publishing opportunity. From the mid-1960s to the early 1970s, dozens of books and collections in a new technology & society genre identified technology as both cause and tool of militarism, environmental degradation, capitalist greed, authoritarian power, sexism and racism, loss of self and community, and impending societal collapse. The team behind *IST* saw a pivot in this new market. What was needed to overcome technocratic and Luddite views, they concluded, was not more traditional scientists and engineers, but rather a new kind of professional who could master the forces of change.

SELF-HELP FOR CREATIVE CAPITALISTS

One of the most impactful and widely read articles on out-of-control technology owed its existence to a shake-up in the trade magazine industry. Paul Goodman's November 1969 *New York Review of Books* essay "Can Technology Be Humane?" diagnosed the technological roots of global unrest and argued for the systemic revision of science and engineering. The enterprise's problems extended beyond military cooptation and the neglect of domestic needs to the very essence of what it meant to be human. Technology, however, if decentralized, collaborative, reflective, and sensitive to the environment could play a critical role in a flourishing democracy. Before it became a touchstone of humanist academics, however, Goodman's article first appeared as "The Case against Technology" alongside an article on "How a Technical Man Can Invest" in the new magazine *Innovation*.[28]

In 1969, Colborn, Maass, and a dozen associates abandoned traditional publishing to form Technology Communication. The start-up's audience was not the scientific enterprise at large, nor even the leading ten percent to which *IST* aspired; the Innovation Group instead would be the elite of true innovators who found themselves at odds with failing corporate and government systems but had the power to change them. Its Innovation Group would be a network in tune with a Heraclitan age and the journalists of *Innovation* would be its facilitators.

The Innovation Group's advisors, most of whom later starred at the Long Island workshop, lived up to this lofty billing. They included Bell Labs' Morton; 3M vice president of research and development Robert Adams; former US congressman and architect of the US Office of Technology Assessment Emilio Daddario; Fairchild Semiconductor chairman C. Lester

Hogan; engineer and policymaker J. Herbert Hollomon; NEC founder Koji Kobayashi; the psychologist Donald Marquis, who founded the technology management program at MIT's Sloan School; Emmanuel G. Mesthene, director of the Harvard Program on Technology and Society; and Gert W. Rathenau, a solid-state physicist at Philips Eindhoven.

In form and content *Innovation* reflected hip exclusivity. Designed by the Madison Avenue firm Chermayeff & Geismar, the magazine had a colorful and clean style that blended modernism with countercultural flourishes. As a special point of pride, *Innovation* was completely member-supported and carried no advertising. The choice of topics cut across disciplines and industries. Informal interviews with contributors and a breezy, conversational tone made even the most esoteric subject seem accessible and familiar. An annual membership fee of $75 included a subscription to *Innovation,* a newsletter, telephone conference calls with authors, a membership database, and invitations to in-person events like the Long Island meeting. Later, the company would offer subscriptions to the magazine for $35, making it one of the most expensive magazines on the planet.[29]

Technology Communication promised its anxious elite an explanation of the underlying forces of societal upheaval, their role in it, and the possibility for managing it. *Innovation's* charter issue was dedicated to *fluidity.* The lead article, "How to Survive a Revolution," by University of Cincinnati president and leadership guru Warren Bennis outlined the tenets of what he called the "Temporary Society." The message was that "as this country heads into the 1970s, it is in a state of such rapid and intricate change that to understand your society you must think of it as a liquid in turbulent flow. You can't rely on solid organizational structures; they'll change. You can't locate any useful boundary between your area of responsibility and the turbulence outside; the outside comes flooding in."[30] The editors were fascinated with the rise of social movements, covering, for example, MIT's 1969 anti–Vietnam War work stoppage. The radicals, *Innovation* argued, had the right attitude toward novelty and had discovered powerful forms of decentralized organization. Editorials also remarked on the transformation in the developed world from overblown praise of science and technology to its outright rejection; "as though technology were spinach," Technology Communication wrote, "and you could like it or hate it."[31]

Technology Communication, in short, was pioneering a new kind of high-tech self-help. The company provided its subscribers with tools for navigating uncertainty via three interconnected strategies. First, it sought to define the traits of a new kind of interdisciplinary change manager. Second, it documented the expert knowledge and practices members would need to achieve success. Lastly, the Innovation Group was to be a social network to build a community of like-minded innovators.

AN ETHIC OF CHANGE

Innovation's dominant theme was the search for identity in a changing world. Like the scientist's ascension in the age of reason, the innovator was a product of their time. Unlike earlier heroic personae of engineers driven by thrift and grit or scientists pursuing pure knowledge, innovators were Heraclitan pragmatists. They accepted the erosion of old ways, were skeptical of fixed positions, and understood the need to act.

Schön told Innovation Group members that society's leaders, managers, scientists, and, indeed, most professionals operated with an ethic of control in pursuit of a "stable state." But the accelerating pace of change had shattered this myth and, along with it, false hopes of a steady progression "toward a good society whose objectives remain stable and clear throughout." Technological disruptions undercut professional skills and structures of belonging that filled practitioners with psychic unease. Schön's solution was to cultivate a way of life "in which transformation around the new is a value in itself."[32]

To thrive in a postindustrial landscape required becoming a new kind of protean man who was guided by an "ethic of change."[33] Perpetual instability demanded continuous self-renewal through experiments in living in "the here and now" and "seeking out the new." An ethic of change was a set of moral virtues that would provide "nuclei for identity and self-worth."[34] Success for these new self-made men was less the result of a priestly duty to objectivity as it was the ability to cut paths across disciplines and institutions in pursuit of one's goals. The blending of technical and social expertise was a prerequisite while collaboration and improvisation were among the most coveted skills. Innovators swam in countercultural currents. They achieved financial and technical goals by drawing on human empathy, they took risks, and, most importantly, they achieved external rewards and inner satisfaction from their faith in the new. This *change manager* was an adaptive agent who would fight bureaucracy with collaborative creativity,

launch entrepreneurial ventures, and save society from both runaway technology and antimodern revolt.

How did Technology Communication personify life in flux? First, theorists like Bennis and Schön described their experiences across corporate,
government, and academic institutions. That picture was augmented
by profiles and interviews with innovators who were to be admired and
emulated.

Innovation constructed intimate portraits of exemplary innovators. In the
September 1971 issue, for example, the magazine published one of the earliest profiles of computer visionary Douglas Engelbart, inventor of the computer mouse, graphical user interface, and pioneer in human–computer
interaction. It identified the Stanford Research Institute researcher with the
antiauthoritarian protagonist of the 1959 short story "The Loneliness of the
Long Distance Runner." Engelbart was "broad-shouldered" and "athletic,"
"wistful and boyish" but "prematurely gray," "diffident yet warm," "gentle
yet stubborn," someone who "wins respect." Personality and physique mattered, as did passion. Engelbart had a "love affair with augmentation systems" that led him to develop groundbreaking technologies, including the
computer mouse.[35] What counted above all was the researcher's vision and
the tenacity to bring it to life. Because Engelbart knew that a "revolution
like the development of writing and the printing press lumped together"
was coming, he had to withdraw from the world in the single-minded pursuit of its transformation.

In *Innovation*'s view, the ideal innovator was a managerial iconoclast who
wasn't focused solely on creating new ventures but could also transform
crumbling infrastructures for a dynamic age. Here, the role of the innovator blended with the identity of organizations. In the November 1970 issue,
psychologist Frieda B. Libaw described how the change manager was an
archetype for "the creative corporation." Libaw, the president of a small
learning sciences business called "Cognitive Systems, Inc.," described the
evolution of corporate organization from the British East India Company
to the present with each era personified by the tools and knowledge of
its time: the new world explorer with his cartographic instruments; the
railroad magnate and his industrial blueprints; the organization man with
his systems diagrams; and, finally, the change manager, whose tool was
networked inquiry. The creative corporation was dedicated to "the solution of social problems," grounded in "behavioral, social and information

sciences," and structured by a value system "that includes antimaterialism and the recognition that individual and community welfare are inseparable." Linking her experiences to the small group technique used by Japan's Sony Corporation, Libaw described Cognitive Systems, Inc. as a "company of activist scholars and scientists" that demanded all its employees be "generalists." According to Libaw, this social entrepreneurship had emerged from the revolt against corporate hierarchy. She concluded that new "participative management" based on humanist philosophies and "social problem solving" were the best hope for reenrolling a generation of disenchanted youth with "constructive and satisfying work."[36]

Innovation chronicled with gusto how a select few individuals and organizations could achieve astonishing levels of creative and financial success as managers of change. These explorations of what made innovators and innovative organizations tick performed multiple functions for authors, subjects, and readers. Authors (whether journalists, engineers, or administrators) adopted the persona of innovation expert. Subjects of their work gained enhanced respectability in tropes that highlighted their evolution from outsiders to thought leaders. Readers gained an aura of exclusivity, insider advice, and a feeling of participation in a dynamic economy with important social goals. They learned that what mattered was not what a man had done, but what he could do next.

MINDSETS AND TOOLKITS

Technology Communication's principal product was access to innovation expertise. In meticulously researched longform essays, Innovation described remedies for lumbering corporations, environmental pollution, and urban decline; it followed new companies throughout their struggles to survive; and it provided primers on the shape of the near future, investigating the leading edge in computing and the development of holography at the intersection of art and engineering.

This was knowledge that could not be found in textbooks, academic journals, or business school cases. Technology Communication's journalists captured the experience and lessons of the entrepreneurs and managers with whom they worked to tell their stories. Meanwhile, applied experts in group psychology, philosophy, landscape architecture, and research management used Innovation to refine their ideas for a practitioner audience.

FIGURE 4.1

Corporations personified as individuals throughout history in *Innovation*. The age of exploration (bottom left), followed by the industrial era (bottom center), the twentieth century's technological corporation (bottom right), and the new "creative corporation" (top).

Technology Communication's promise was that by concentrating and distilling this expertise, the start-up could equip its customers with tools for mastering innovation.

The innovator's most powerful tool was their ethic of change. However, to succeed, the innovator's mindset needed to be in tune with the shifting external environment. In his 1969 article, "The Diffusion of Innovation," Schön replaced a linear model with one in which innovators acted in a complex, dynamic, and evolving network. This distributed agency could be seen in military–industrial systems and social welfare programs but was especially characteristic of radical politics. In "The Movement," Schön wrote: "Orthodox Marxism, the theory of power elites, the radical sociology of an Alinsky, the radical critique of a Chomsky, the doctrines of participation and advocacy, the Black Manifesto, the elaborations of guerrilla ideology and tactics, the rationale for mysticism and for the use of drugs, the philosophical radicalism of Marcuse, the essays in radical politics of the new left, all flow and work together and change rapidly over time." Amid loose and evolving connections, "centers come and go and messages emerge, rise, and fall."[37] Schön championed these organizational modes and argued that they were essential to the changing nature of science, technology, and business.

To illuminate the patterns of change, *Innovation* provided a forum for managers in the automobile, biomedicine, and aerospace industries. Most of Technology Communication's lessons, however, addressed the digital age. *Innovation* covered microelectronics, computers, information services, and more. An exploration of the time-sharing business, for example, documented how entrepreneurs were building the platform for what would become the internet and analyzed the team composition of promising start-ups. Another explored why regulating technological change was so difficult by diving into the challenges faced by the Federal Communications Commission, an organization originally created to address telegraph exchanges.

Innovation portrayed digital technology as the driving force of innovation culture. Its articles provided the first draft of the history of the computer industry, one on which virtually every subsequent account of Silicon Valley is based. A 1970 interview with Robert Noyce and Gordon Moore, cofounders of Fairchild Semiconductor and Intel, is illustrative. "Building a Rational Two-Headed Monster" portrayed Noyce as a warm, but deeply competitive leader, recounting a ping-pong game in which the five-year-old

Noyce got angry when his father let him win. "If you're going to play, play to win."[38]

Technology Communication also established innovation expertise as a universal toolkit. In 1971, the company helped Morton produce the textbook *Organizing for Innovation*. Working with journalist David Allison, who had edited his 1964 *IST* article on the innovation process at Bell Labs, Morton described how the systems approach provided the means for overseeing innovation as it changed the industry that spawned it. The systems approach, however, had shifted from engineering to "a *new viewpoint* and a powerful *method*" of the "ecological-systems view." Drawing on intellectuals like his Innovation Group friend Bennis, Morton argued that change management applied to "posing and solving problems . . . whatever their purpose and environment." *Organizing for Innovation* captured Technology Communication's ambition to "resolve the apparent conflicts between business and science, between technology and society, and between the individual and 'the system.'"[39]

By 1972, Maass had synthesized the Innovation Group's findings into a "self-energized program" for training change managers in any organization. The program started by guiding these managers through a clear-eyed analysis of technological change within their organizations and industries, informed by management science. Local innovation groups of twelve to twenty change managers could then formulate and execute "action plans" in their given organizations. Technology Communication facilitated the process by providing "discussion packages" and a library of then-cutting-edge videotapes. These local innovation groups would work to cultivate institutional "climates" that were "flexible, responsive, and innovative."[40]

COMMUNITIES OF PRACTICE

A key difference between Technology Communication's innovation expertise and classic self-help programs was its commitment to building a community. Staff saw themselves as "environmental creators," describing *Innovation* "not so much as a magazine as a vehicle for initiating a process of interaction." Technology Communication extended the boundaries of print media in search of new ways to connect people and ideas via a social technology that it hoped would spawn innovation groups across the world. A kind of proto-LinkedIn, *Innovation* was intended as a space for sharing problems, putting seemingly disparate ideas on an equal plane, and aiding entrepreneurs in their pursuits. The July 1969 issue, for instance, offered members a direct phone line to venture capitalists Georges Doriot and Arthur Rock. Likewise, in tandem with *Innovation*'s profile of Engelbart, members could sign up for an in-person seminar with the computer visionary in his laboratory at the Stanford Research Institute, where they could try out the augmentation system made famous by his 1968 public demonstration.[41]

Within its first year, the Innovation Group had 5,000 aspiring change managers, with Maass aiming for an upper bound of 25,000. Subscribers consisted of executives, staff engineers, senior scientists, and professors from Lockheed to the National Academy of Sciences and the University of Southern California. Colborn described *Innovation*'s readership as a "zestful crew who mostly felt good about the state of their lives" in contrast to popular opinion which cast "engineers and applied scientists are the most discontented professionals in America."[42]

In January 1972, Technology Communication expanded its electronic services by partnering with another start-up called "TeleSession Corp." Founded in 1970 by marketing expert Ron Richards and psychologist George Silverman, TeleSession started as a late-night telephone-based chat group, where strangers could discuss common interests. The Innovation Group used the TeleSession switchboards to build virtual workshops around its articles. These spaces offered members a place to perform the rituals and established shared identities of being innovators. Branding itself as "the first magazine with an interactive feedback system," *Innovation* boasted that economist Milton Friedman dialed into TeleSession from his summer cabin and deemed the experience more valuable than a face-to-face meeting. "Complete strangers," Technology Communication enticed, "end up talking as though they've known each other for years—and wanting to call on each other's expertise again."[43]

In addition to forging direct connections among members, Technology Communication analyzed the culture it was facilitating. Here again, the media start-up gave life to dominant myths of the digital age. The December 1969 issue of *Innovation* published "The Splintering of the Solid-State Electronics Industry," one of the most influential articles ever published about Silicon Valley. Reporter Nilo Lindgren, with the help of local journalist Marion Lowenstein, described the birth, evolution, and futures of

"Semiconductor Country." Using a genealogical chart, the article built on prior work of *IST* and *Innovation* contributors to document the central role of Fairchild Semiconductor in the making of Silicon Valley. Lindgren recounted the story of William Shockley's departure from Bell Labs after inventing the transistor to start the West Coast semiconductor industry, the eight engineers who abandoned Shockley to start Fairchild, and the spinoffs that followed. Lindgren argued that the pattern of growth in the Bay Area represented a "fundamental change" in science from an academic to an entrepreneurial pursuit. It wasn't just the quest for profit, pleasant weather, or advancing technology that explained Semiconductor Country, however, it was the region's "character."

Lindgren provided a map of Semiconductor Country to demonstrate the clustering and explained that regional growth was reproducing the "smog" and "hustle and bustle" that its East Coast transplants had hoped to escape. A detailed look at twelve start-ups, from Advanced Micro Devices (AMD) to Qualidyne Corporation, accompanied by descriptions of forty different founders and their career trajectories speculated on which spinoffs were likely to succeed and which to fail. Not all was sunny, however, with one insider describing many start-ups as "parasites" seeking public funding so that founders could "sell out" and leave "others holding the bag." Throughout, Lindgren analyzed the complex set of climatological, technological, and organizational factors but concluded with its culture and values. The "persuasive aura-of-well being," *Innovation* concluded, was due to "the atmosphere of mutual help that pervades the peninsula."[44]

The trope of Silicon Valley as an innovator's utopia is so well-worn that we forget it was once novel. *Innovation* established the narrative. First, its citizens identified the alignment of technoscientific and economic success as their highest purpose. Science that couldn't make it from lab to market was impotent, while pure profit-seekers were parasites and sell-outs. Second, it replaced the dehumanizing hierarchies of cold warriors and corporate executives with collaborative exchange and the blurring of work and social life. The region's "enormous camaraderie" existed "despite their being competitive on a business basis, and their business willingness 'to wipe someone out,'" because "the high technology people mix frequently socially, engage ideas, and have a genuine high regard for each other." Third, it replaced the drafting room, lab bench, and lonely office with pool parties, surfing, and drinking at the Wagon Wheel saloon, while at the same time

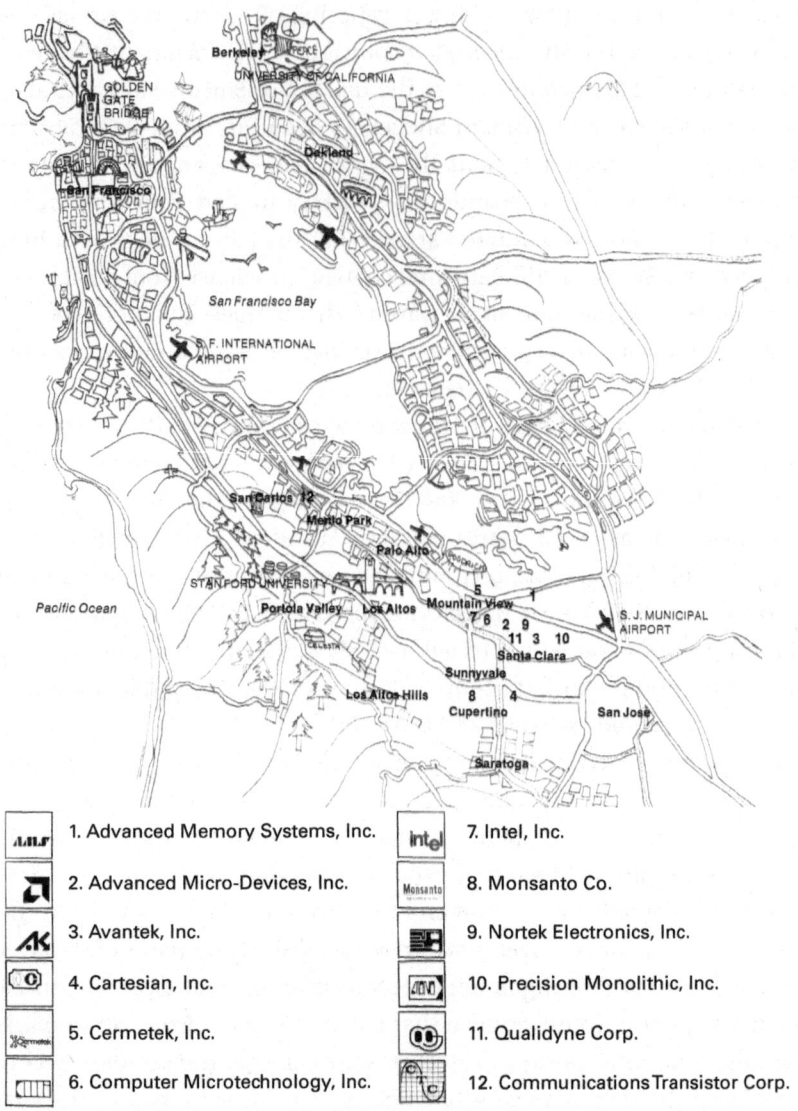

ΛIIf	1. Advanced Memory Systems, Inc.	intel	7. Intel, Inc.
	2. Advanced Micro-Devices, Inc.	Monsanto	8. Monsanto Co.
ΛK	3. Avantek, Inc.		9. Nortek Electronics, Inc.
	4. Cartesian, Inc.		10. Precision Monolithic, Inc.
	5. Cermetek, Inc.		11. Qualidyne Corp.
	6. Computer Microtechnology, Inc.		12. Communications Transistor Corp.

FIGURE 4.2

Map of "Semiconductor Country." Published in 1969, this was the first of many similar illustrations of Silicon Valley that spatially located the area's technology companies. See figure 1.2 for a recent example of this approach.

championing an intimate family community. Fourth, *Innovation* portrayed that community as international, diverse in its personalities and its pathways to success. Finally, this community was a participatory "open-ended proposition" characterized by an instability that was risky but exhilarating. "It is almost enough," Lindgren opined, "to make an Easterner bolt for the semiconductor paradise, especially as winter dulls the skyline and icy slush impedes progress along the sidewalks."[45]

FROM CHANGE MANAGERS TO INNOVATION MILLIONAIRES

As today's self-help guides for innovators warn, most ventures fail. Sometimes, the difficulty is transitioning from lab to market; at other times, a poor business plan, a soured partnership or simple bad luck. Technology Communication's demise was marked by personal tragedy and financial trouble. Just nine months after the Long Island workshop, Colborn died of cancer at the age of fifty-nine. Then, in 1971, Morton, known for his hard-charging lifestyle, arrived at a New Jersey bar after closing. He encountered two would-be muggers and a fistfight ensued. Morton was beaten, put in his car, and set ablaze in a gruesome murder. Compounding personal tragedy, a post-Vietnam recession in the defense industry thwarted the company's expansion. In 1972, the magazine unceremoniously published its final issue.

The core themes of *Innovation*, however, exploded in importance in the 1970s as the tectonic shifts of the post-Vietnam era became clear. There was a booming market for magazines, books, and corporate ephemera that imitated *Innovation* in style and content. These included corporate magazines like DuPont's own *Innovation*, articles and special supplements in business media such as *Fortune*, science and technology outlets such as *IEEE Spectrum* and *Technology Review*, and West Coast trade papers like *Electronic News*. Former *Innovation* contributors contributed to this proliferation. Its staff dispersed to technology publications and corporate public relations and its advisory members remained prominent thought leaders. Maass, the surviving member of the founding partners, returned to mainstream publishing to head *Electronic Design* during the microcomputer revolution.

The growth of innovation publishing in the early 1970s closely followed Technology Communication's blueprint but extended it to a wider audience. Throughout the 1970s, books and magazines dramatized scientific entrepreneurs as the nation's most important asset. Managerial guides to creativity, entrepreneurship, and organizational change based on Morton's and Schön's models likewise flourished.[46] These texts created a model of change management that constitutes a major share of recent books like *The Innovator's Toolkit* and *The Innovator's Cookbook*.[47] Likewise, we can see the legacy of Technology Communication's multimedia experiments in TED Talks and executive immersion seminars.

As more and more media outlets promoted innovators, the virtues of the change manager remained, but they were directed toward full-throated entrepreneurship. Take, for example, Gene Bylinsky's bestselling *The Innovation Millionaires*. Bylinsky was a technology writer for *Fortune* who established himself as a computer-industry journalist after Technology Communication's breakup. The *Innovation Millionaires* began with a nod to Cold War science with a foreword by Manhattan Project physicist turned nuclear consultant, Ralph Lapp. Lapp railed against decades of military–industrial policy and in contrast championed innovator–entrepreneurs as Magellans who "set out upon uncharted seas on subtle voyages of discovery where monetary rewards are more than matched by the immense payoff in human benefits."[48]

The Innovation Millionaires offered biographies of entrepreneurial innovators, all men, that emphasized their traits and mapped the innovation process through their personal stories. An interview with DEC founder Kenneth Olsen was a history of Boston, the Cold War tech boom, and its bust in the early 1970s. Olsen provided an insider's look at the downturn optimistically, arguing that the post-defense environment presented more healthy markets and that a recession was the best time to become an entrepreneur. These profiles blended Bylinsky's voice as an expert with quotes from his subjects. They provided a history of innovation, touched upon failure, risk-taking, passion, and analytical intelligence amid "a hurricane of technological change." In a profile of Georges Doriot, for example, readers learned about the origins of venture capital as a model for funding innovation through a folksy account of how prior to scoring multimillion-dollar profits—Doriot made just $87 his first year. They learned Doriot's axioms—watch the cocktails—and read his successor's conclusion about how his approach to venture capital in 1950s Boston had become systematized by Arthur Rock and others in California.[49]

Bylinsky's portrayals of innovation millionaires were morality tales that demonstrated how easily the innovator's persona could extend from microelectronics to other fields such as biotechnology. These profiles linked technoscientific entrepreneurship to "improving lives." Alejandro Zaffaroni, for example, built a drug delivery system company based on collaboration between medical researchers, scientists, and engineers. The result utilized the body's natural molecules to reduce the harmful side effects of drugs and led to products used to tackle chronic diseases in the underdeveloped world.[50]

In the 1970s, innovation popularizers also completed the migration from East Coast to West Coast, with a cohort of Bay Area journalists adopting a similar approach to *Innovation*. Don C. Hoefler, for example, became famous for coining the term "Silicon Valley" in a series of *Electronic News* articles that rehashed Lindgren's "Splintering of the Solid State" for a local audience.[51] Meanwhile, traditional newspapers, most notably the *San Jose Mercury News* cultivated technology journalists adept at both the promotional puff piece and the longform expose.

While most early Silicon Valley innovation literature focused on what it took to succeed, the desire for community was also a constant theme. Hoefler sang the praises of a culture where "the wives all know each other and remain on the friendliest of terms. The men eat at the same restaurants; drink at the same bars, and go to the same parties." This new region anchored cosmopolitan innovators in a parochial home. "Despite their fierce competition during business hours," wrote Hoefler, "away from the office they remain the greatest of friends."[52]

In the early 1980s, the personal computing industry sparked a mass media discovery of Silicon Valley and the digital innovator. *Life*, *National Geographic*, *Time*, and television news championed the innovator's persona.[53] This burst of attention again emulated Technology Communication's model. *Newsweek*'s special issue *Access: The Magazine of Life and Technology*, for example, copied the strategy of exclusive insight. *Newsweek* boasted that "this singular effort is being sent to a select group," not innovators themselves, but rather the early adopters of the digital technologies they produced.[54] Its coverage featured in-depth personal stories including excerpts from Steven Levy's soon-to-be bestseller *Hackers* and an interview between tech reporter Tom Zito and a young Steve Jobs. Other mass media accounts celebrated and gently ribbed the unique culture of innovators, describing offices in which "video games abound, ping-pong tables are in use, speakers blare out music ranging from the Rolling Stones to Windham

Hill jazz. Conferences are named after Da Vinci and Picasso, and snack-room refrigerators are stocked with fresh carrot, apple, and orange juice. (The Mac team alone spends $100,000 on fresh orange juice per year)."[55]

SCRIPTS FOR LIVING

By the mid-1970s, the persona of the innovator had coalesced in a different form than the civic-minded prototypes of the 1960s. This identity was constructed by a group of entrepreneurs in the publishing industry who combined image-making, the communication of expertise, and the building of social networks. These journalists were just as important as academics, scientists, and policy makers in shaping innovation; indeed, they provided a publishing platform for these technology managers to reinvent themselves as innovation gurus. As Silicon Valley emerged as a national power center, these journalists also provided foundational narratives that defined the region as an innovative *community*. The effect was to clarify and reinforce a set of beliefs about historical change and the people responsible for it.

Profiles of the innovator as change manager had a three-part thesis: innovators exist and want to be understood, innovators established new ways of living, and the values of those ways of living were a social good to be cultivated. Technology Communication went further, providing spaces for members of the Innovation Group to perform innovative lives.

Nonetheless, being an innovation millionaire had its downsides. There were constant financial risks, the challenges of satisfying diverse stakeholders, and the existential anxiety of living with permanent instability. Staying in character could also get tiresome. Take Jobs's interview with *Newsweek*'s reporter Zito. Frustrated about being turned into a stereotype, Jobs snapped when asked yet again to square his Eastern mysticism with his billion-dollar company: "Yeah, well, OK. What do you want me to say? Give me the possible responses, I'll pick one."[56]

The scriptwriters also could become jaded. For example, *San Jose Mercury News* technology reporter Michael S. Malone borrowed from Thomas Hobbes's *Leviathan* when mapping the innovator's community in "The 100 most powerful people in Silicon Valley." At the center of power, were the *immortals* like Packard, Hewlett, and Noyce. They were followed by young *demigods* like Jobs. The community was also populated by venture capitalist *bagmen*, futurist *gurus*, public relations *flacks*, local *pols*, real-estate *pick-pockets*, environmentalist *gadflies*, and, of course, journalist *hacks*.

A second major concern was how innovation self-help could become truly collective. Technology Communication had built community among an insider elite, and high-tech spinoffs in Boston and Silicon Valley offered on-the-ground education. But how could innovators learn their art formally and on a wider scale? In the 1970s, universities across the United States looked to establish teachers, precedents, and tradition.

5 INCUBATING ENTREPRENEURS

Because we don't know all about embryology doesn't mean we don't have chickens.
—Dwight M. Baumann

Imagine a point-to-point transportation service in which two parties communicate at a distance. A passenger in need of a ride contacts the service via phone. A complex algorithm based on time, distance, and volume informs both passenger and driver of the journey's cost before it begins. This novel business plan promises efficient service and lower costs. It has the potential to disrupt an overregulated taxi monopoly in cities across the country. Its enhanced transparency may even reduce racial discrimination by preestablishing pickups regardless of race.

Prototyped in 1975, the automated taxi dispatch system was the brainchild of mechanical engineer Dwight Baumann and the students at Carnegie Mellon University's (CMU) Center for Entrepreneurial Development (CED). Designed to resurrect the Peoples Cab Company, an insolvent business that once served Pittsburgh's African American neighborhoods, the dial-a-ride project combined high technology, creative students, and expert mentorship for urban renewal.[1]

CED was part of a novel federally funded experiment to remake American science to serve national needs. Faculty in engineering, business management, and urban and public affairs collaborated to develop innovators in the experimental "hatchery."[2] Their pursuit of market-based technological solutions to social problems challenged the norms of research science and higher education as they fostered risk-taking and resilience and birth campus start-ups.

Today, university incubators are no longer experiments. Hundreds of such centers seek to nurture the next Uber, but also social ventures like the Peoples Cab Company.[3] These initiatives aim to transform ideas into businesses, discoveries into applications, classroom assignments into university revenue, and faculty and students into entrepreneurs. The idea that universities are engines of innovation is so ingrained in the fabric of university life that we take it for granted that it was always the case.

University incubators are sites of promise but also controversy. Practical concerns include how ought venture incubation impact universities' tax-exempt status? Who owns intellectual property resulting from the work of students and faculty? For what kinds of activities should students earn academic credit? But innovation centers raise deeper quandaries about the university's soul. Where boosters view innovation centers as a means of applying research to human needs, critics identify a neoliberal market logic with corrosive effects on knowledge itself. These questions and tensions about intellectual value, public versus private benefits, and the proper ends of academic inquiry long predate both current innovation centers and the experiments at CED, but their answers and configurations have changed and evolved.[4]

This chapter returns to the moment when innovation incubators were themselves incubated. To explore how and why universities came to grapple with the practical and moral questions of innovation within the university, it investigates the first formal efforts by academic scientists and engineers to train innovators. Between 1973 and 1978, the National Science Foundation (NSF) awarded million-dollar grants to CMU and three other schools across the country. These universities conducted a series of experiments on how to apply the fruits of academic research to better serve national needs. Underlying the Center experiment was the belief that fostering innovation went beyond policy levers like economic incentives and patent law; it was a task of creating new kinds of people.

CAN INNOVATORS BE MADE?

Higher education's embrace of entrepreneurship resulted from a combination of top-down and bottom-up efforts to reform the Cold War university.

Observers often identify the early 1980s as the beginning of this transformation and attribute the rise of campus innovation centers to neoliberal ideology and the need for revenue to sustain university budgets. Universities had expanded dramatically in the postwar era due to federal funding. In the 1970s, when the influx of funding leveled off, universities sought new income sources. At the same time, large corporations had also begun to cut their research divisions and looked to universities as sites of profitable ideas. In this explanation, the Bayh–Dole Act of 1980—which granted companies and universities patent rights to discoveries made with federal funding—stands as the catalyzing event in a new regime of technology transfer and science commercialization. However, the federal government began to foster market logic well before Bayh-Dole.[5] By exploring the earlier experiments in innovation education, I recreate the on-the-ground interactions among students, faculty, NSF program officers, and the technologies and ventures they developed.

When the NSF conceived of the Innovation Center Experiment in 1972, the operational model for training scientists and engineers in the United States was one of manpower in service to a linear model of innovation. For nearly three decades, scientific and technical education was shaped by the ideal of scientists pursuing "basic" discovery in universities and federal laboratories, engineer–scientists conducting "applied" research elsewhere on campus, engineers developing those ideas in giant teams for companies such as Lockheed and Boeing, and research managers overseeing the process. This system formed the basis of the NSF, dictated national science policy, elevated the scientist as a national hero in pursuit of truth beyond politics, and pumped hundreds of millions of dollars into higher education.[6] In practice, the lines between basic and applied research were blurred, but the perceived hierarchy was ingrained in the NSF and the university research culture that it helped to foster.

In the late 1960s, the postwar system of academic science and engineering appeared to be breaking down for both material and ideological reasons. Science and technology increasingly were seen as root causes of environmental destruction, the Vietnam War, job losses, and racial and economic inequality. Universities were simultaneously major partners in the military–industrial complex and the source of resistance against them in the form of student and faculty protesters. At the same time, the microelectronics industry and the rise of Silicon Valley were changing the nature of technoscientific labor and diverting talent from the academic ranks.[7] A similar reckoning took place at the level of national science policy, where critics on the left attacked the complicity of scientists in the military–industrial complex and those on the right assailed the wastefulness of ivory tower spending on science.

In this moment of revolt, a bipartisan vision of innovation had begun to take shape among federal bureaucrats as a means of reconciling the challenges facing the United States. The Department of Commerce's 1967 Charpie Report came out of Great Society boosterism for innovation as a driver of progress through consumer technologies like television, air travel, and microelectronics.[8] The Nixon Administration embraced these concepts as a solution to meeting economic challenges from Japan and Western Europe, putting unemployed aerospace engineers back to work, balancing environmental impacts of technology, and rejuvenating American cities.

Innovation experts in Washington, DC, and the booming technology regions of California and Massachusetts had begun to promote innovators as the people who would bring these changes about. A decade studying innovators in environments as varied as native tribes and scientific laboratories convinced experts that these change agents were different from the established leaders of American science. Their propensity for venturesomeness put them at odds with the university seminars and corporate bureaucracy that characterized scientific work in the military–industrial complex. The Charpie Report enshrined entrepreneurial innovators such as Polaroid's Edwin Land and Texas Instruments' Jack Kilby as drivers of high technology. Meanwhile, the microelectronics industry was changing the high-tech economy and the norms of what a career in science and engineering could look like. Journalists worked with entrepreneurs, engineers, and scientists to forge an image of innovators as entrepreneurial change managers in a global economy who could tackle large social and technological challenges.

The question for government, industry, and academia was how such scientific and technological innovators could be made. Instead of producing the "same class of men" to be interchangeable, the goal was to create new kinds of people capable of building and thriving in a flexible market system.[9] An even more daunting problem was whether universities were willing to confront the institutional task of remaking themselves.

Innovation's champions saw some positive signs for innovator initiatives in the 1960s and early 1970s. Business schools were starting to pursue entrepreneurship education with an emphasis on science and technology, not just at Stanford and MIT, but at public and private research universities across the United States.[10] Expert venues such as *Industrial Research* added empirical and theoretical grounding to the patterns of academic spin offs and corporate "incubator organizations."[11] Innovation advocates also established a series of programs to aid would-be entrepreneurs. Milwaukee, Wisconsin's new Center for Venture Management, for example, compiled a handbook that listed nine hundred local, regional, and national resources for aspiring entrepreneurs.[12]

The roadmap for transforming higher education to support innovation originated from the bureaucratic innovators of the Kennedy and Johnson administrations. In 1966, J. Herbert Hollomon and his Department of Commerce team of Daniel De Simone and Donald Schön collaborated with the NSF to host an Education for Innovation convening that brought inventors from industry and academia together with advocates in government to explore how higher education could support "creative" engineering education. The group concluded that engineering students were future "manipulators of change," and that universities needed to learn how to develop and direct students to technical and social issues of national importance, including transportation, housing, and communication. Participants bemoaned that the cult of science and its emphasis on rigor and mathematical proofs killed creativity. Students could solve predefined puzzles, but they could not design for ambiguous and open-ended situations. Likewise, the educators and policymakers described an almost total absence of training in technology ventures such as appraising new ideas, understanding the dynamics of industry, the "interplay" of technology and social trends, and the securing of venture capital. To change university culture, the group proposed increased cooperation between universities, industry, and government. Participants lamented that university culture looked down upon creativity and entrepreneurship as subordinate activities and suggested that in this hostile environment, the best way to support entrepreneurial students was to "just love them."[13]

To simplify a complex story, bureaucrats in both Democratic and Republican administrations, independent inventors, experts in fields from anthropology to economics, business leaders, and the engineering profession had come to identify innovators as agents of national progress. They also concluded that these innovators could be taught in the nation's universities. In the 1970s, it fell upon the NSF to develop successful models for producing these risk-taking sociotechnologists.

THE INNOVATION CENTER EXPERIMENT

The University Innovation Centers were part of the NSF's broader reconciliation with the barrier between basic and applied science. During the Nixon administration, NSF director H. Guyford Stever insisted that the agency explore new directions. Simply emulating the Stanford and MIT models of sponsored research was not a viable path. It was expensive, difficult to emulate, and politically problematic.[14]

In 1972, shortly after Nixon announced his Presidential Prize for Innovation, Stever established the Experimental R&D Incentives Program. The initiative was to "incentivize" innovation for national needs by conducting research on "how the government [could] most effectively accelerate the transfer of new technology into productive enterprise."[15] Experiments ranged from fostering industry–university cooperation to developing medical technologies where the market failed.

Experimental R&D Incentives was controversial from the start. Not only did the program amplify debates about the nature and purposes of scientific knowledge, but it also complicated them by bringing corporate managers into the NSF. It also set an agency that shaped its identity around "pure" science on a path of "socio-technical" research.[16] Stever appointed C. B. Smith to lead the initiative, a research manager at United Aircraft Corporation who lacked a PhD. Smith, in turn, added engineers with industrial experience to his staff, including Robert Colton, an automotive engineer, with responsibility for the University Innovation Center Experiment. Stever stressed the experimental nature of the project because many in the NSF and the scientific community resisted goal-directed research. Innovation, with its connotations of profit and social change, was even more suspect. Rather than a full-on mandate, the softer pitch for Experimental R&D Incentives was as an empirical test of innovation expertise, including whether such expertise could be taught.

The Innovation Center Experiment attempted to operationalize the range of ideas about innovators and entrepreneurs developed at the Department of Commerce's Woods Hole meeting. Centers would have three main functions: first, to educate students in innovation and entrepreneurial

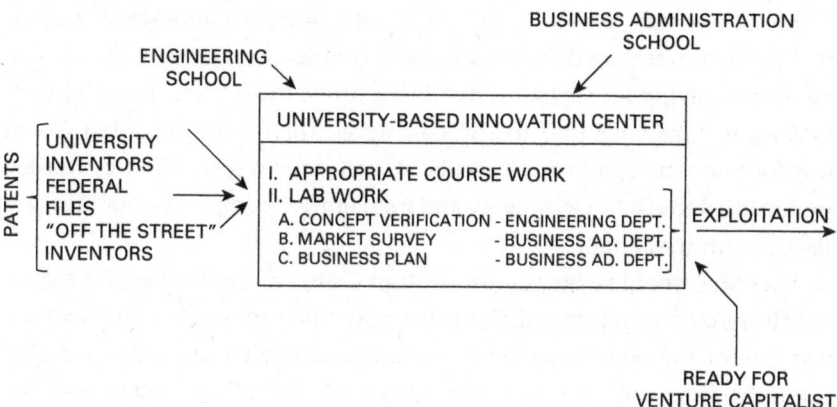

FIGURE 5.1
Design for the Innovation Center Experiment (redrawn from NSF original).

processes; second, to research and improve the innovation process; and third, to support independent inventors in bringing ideas to market. Alongside these motivations were a set of operational goals. Ideally, the Centers would generate successful entrepreneurs, accelerate the development of technology from university research, and, ultimately, pay for themselves.[17]

The NSF chose four universities distributed across the country to capture a range of approaches to innovation incubation at public and private schools with different missions: MIT, the University of Oregon, CMU, and the University of Utah. MIT targeted undergraduate students through formal coursework and an innovation "co-op" that assisted in turning ideas into products, Oregon evaluated the ideas of garage inventors from across the country, CMU established a nonprofit corporation to support graduate student ventures, and the University of Utah emphasized an ecosystem of faculty and student biotech and computer graphics start-ups coming out of its research labs.

The program's definition of *experiment* was more aspirational than rigorous. The theory being tested was that centers might fail, or succeed but make little broader impact, or prove the model's value. The experiment was to give the centers money and see what happened.[18]

MIT: ENTREPRENEURS FOR SOCIAL NEEDS

"Are you an innovator?" a pamphlet with a white dove flying against a murder of crows asked MIT's undergraduate students in the fall 1973 semester.

If the answer was "maybe," then MIT's new Socio-Technological Innovation Seminar was your destination. The introductory class would teach you "problem-solving for social needs." If the answer was "yes!" and you were looking to take your idea to the next level, then the Innovation Co-op was for you. There, a team of faculty experts would give you the "needed professional touch in technology and marketing" to improve your idea and secure venture capital.[19]

That MIT would receive an Innovation Center award seemed a foregone conclusion. MIT epitomized Cold War scientific innovation. The institute was one of the largest university recipient of federal research contracts. Administrators and their former students advised US presidents, and the NSF director himself was an alum. Other alumni headed companies along Boston's Route 128 and around the world. The MIT ecosystem had produced gyroscopic missile guidance, radar, atomic energy, space flight, and digital computing.

Why would MIT need to improve its innovation education? Despite the institute's preeminent standing, MIT was in the middle of a major cultural reevaluation. In the late 1960s, critics had assailed MIT as the military–industrial complex incarnate. Its students and faculty had gone on strike in 1968 to redirect research from military applications to societal needs. Undergraduate students also decried a soulless culture which prioritized graduate education and scientific rigor that was hostile to the interests of young people.[20]

Still, the Innovation Center award came as a surprise to almost everyone. Aeronautics professor Yao Tzu Li, the center's director, put the proposal together without MIT support. Li was not part of MIT's inner circle, nor was he connected to the expanding group of business-school innovation experts setting up shop at the institute. He was one of the very few native Chinese faculty members at MIT and in American higher education in general. Li had earned both his masters and ScD degrees in Aeronautics at MIT in the 1930s, but returned to China to lead its aeronautics industry, including the design and construction of an underground aircraft production plant during World War II. Li returned to MIT in 1947 as a research associate on defense projects, including the B-52 bomber. In 1961, after becoming a tenured professor, Li served as director of the Man-Vehicle Laboratory, which studied space sickness among Apollo astronauts. He also created two businesses and received over thirty-seven patents.[21]

Li viewed innovator education as a means for redirecting MIT's military–industrial research to serve civilian economic growth. He argued that moving away from giant government-funded companies like Lockheed required a diverse economic system with "workers who have stronger inclination of seeking out new activities, willing to take more risk, and more eager to innovate."[22] Li believed that MIT lacked these creative risk-takers and that its undergraduates were clamoring for careers beyond the Cold War defense culture.

MIT's Innovation Center had two components. The first was a suite of courses that replaced mathematical proofs with "social needs and problem finding." A Socio-Technical Innovation Seminar anchored the curriculum, utilizing a case study approach that required students to formulate a business plan and explained legal and financial aspects of entrepreneurship. The second was the Co-op, a product-development laboratory that was a space for students to practice, fail, and perhaps even make money. Li imagined the Co-op as akin to a teaching hospital with medical interns. Both components stressed cultivating entrepreneurial mindsets in a "cheerleader" environment.[23] Most students, Li argued, would make the important discovery that they were not destined to be entrepreneurs; he estimated that just 5 percent had what it took. In addition to nurturing the rare few entrepreneurs, the Center would benefit hundreds of other students destined for "large government or industrial organizations" by broadening their horizons and letting them "know what's going on in the world."[24]

The Co-op planned to sustain itself with revenue from its successful entrepreneurs. An advisory board of faculty and industry participants chose projects in which MIT would invest up to $40,000. To receive funding, students required a business plan that included disclosure of essential concepts, a working model, a patent position, competitive position, and a market survey. Patents were front and center. Students, faculty, and staff kept "written records, properly witnessed, of any ideas or concepts they present in the course of their activities."[25] The Center had a novel intellectual property arrangement to protect students' intellectual property but also to maximize the taxpayer's return on investment. Students without external sponsorship would have full patent rights, but by utilizing the Co-op's services, they would give the government a royalty-free license to their product or idea. Projects reaching an advanced stage were encouraged to seek venture capital, at which point the Center would have "sole right" to the invention.[26]

The Center produced a range of companies and successful patent licenses. These included a high-efficiency bow for archery, an earthquake seismograph, and an electric guitar with built-in tone and volume control. The Center's poster child was Philip Doucet, a master's student in the Sloan School. His Computer Controls Corporation developed what we now call a "smart home device," a low-cost thermostat system for reducing energy use by monitoring building temperature.[27]

The Center, however, came under internal criticism from a rival approach to entrepreneurship even before NSF funding arrived. Li's project was one of two nascent innovation programs at MIT established just months apart. The other, the Experimental Development Foundation, was housed in the Sloan School of Management and instead of undergraduate programs, it sought to create entrepreneurial spin-off companies out of faculty research. The Foundation was headed by MIT alumni Richard Morse (BS 1933), the inventor of freeze-dried orange juice and cofounder of Minute Maid, who had refashioned himself as an innovation expert. The multimillionaire entrepreneur returned to his alma mater in 1970 to build an internal venture capital "catalyst." MIT would make small initial investments of $50,000 in faculty spin-offs with the goal of stock ownership in the new companies. To oversee and balance these different visions of entrepreneurship, MIT created an Innovation Education Council led by former Department of Commerce experts Hollomon and Schön who had also settled at MIT. This administrative group diminished the focus of innovation on undergraduates in favor of faculty research.[28]

Once the Center began operations, it was plagued with additional challenges. Most MIT faculty considered product design to be a second-class activity and believed that the Center with its tiny budget was doing frivolous work not up to MIT's standards. Turf battles over which experts owned innovation also prevented the Center from gaining internal support.

Finally, attempts to bring student ideas to market resulted in ethical, financial, and legal quandaries. For example, in 1975, a team of five electrical engineering students created a video tennis game system. MIT patented the product and licensed it to a local electronics firm, which made lofty promises and rushed the product to market. However, the company was unsuccessful in the nascent but fiercely competitive market, and when it

FIGURE 5.2
Student innovator–entrepreneurs from the MIT Innovation Center with their video tennis game. Photo by Calvin Campbell. Courtesy of MIT Museum.

went bankrupt, the Center lost $100,000 and the students were not paid the $15,000 in royalties that they were promised.[29]

One critique that the Center did not encounter was that its activities were a corruption of mission or academic purpose. On the contrary, its sociotechnical problem-solving appeared to align with the goals of MIT's internal and external critics working to dismantle the military–industrial complex.

When NSF funding ran out, Li was so frustrated by his marginalization at MIT that he retired and attempted to create his own external Enterprise Formation Laboratory. David Jansson, an untenured aerospace professor, took Li's place as director. The Center also shifted away from venture development to meet the needs of existing technology firms.[30] Perhaps there was a place for the video game architect, the music technologist, and the environmental designer at MIT. But, in the 1970s, the Institute remained uneasy about how entrepreneurial its students and faculty ought to be.

OREGON: A ROLE FOR TINKERERS

In comparison to MIT, the University of Oregon was a much less likely home for innovator–entrepreneurs. The university did not even have a college of engineering. Rather than an urban high-tech cluster, Oregon's campus was in rural Eugene. And compared to Massachusetts, the state had little venture capital to speak of. Instead of cultivating students, its Experimental Center for the Advancement of Innovation and Invention addressed a different set of aspiring innovators who were struggling on the margins of both industry and academia—America's independent "garage" inventors.

The idea for Oregon's Innovation Center emerged out of a local network of inventors, professors, and civic leaders called the "Oregon Inventors Council" (OIC) that was modeled on the National Inventors Council. OIC's mission was to foster connections among local inventors and small businesses, and, in 1970, it partnered with University of Oregon business professor Leslie L. D. Shaffer to evaluate the ideas of local inventors. OIC saw the NSF experiment as an opportunity to scale its work to the national level by addressing the underserved "cross section of society [that] includes the tinkerer, the housewife, or anyone else capable of creative and innovative problem solving."[31]

The Oregon Center targeted two features of the innovation enterprise that the federal government identified as critical. First, it fulfilled the NSF's desire to evaluate potential innovations at the earliest stage, thus saving money by placing better bets. Second, it tested whether there was still a role for independents, who some government champions valorized as underappreciated drivers of economic growth and the American spirit.[32]

Independents bet their livelihoods on their inventions but had severe knowledge gaps about the financial and legal dimensions of innovation. Thus, they were often victims of predatory "idea brokers" who charged thousands of dollars to evaluate inventions and file patents. To close that knowledge gap, Oregon's Innovation Center disseminated information on the innovation process by emulating agricultural extension stations, evaluating ideas as a public service at no charge to all comers.[33] The Center initially aimed to aid clients face-to-face, but it was overwhelmed by applicants.[34] Internal surveys revealed that only 5 percent of would-be innovators left satisfied with their experience. The Center briefly entered a second phase of operations, during which potential clients received an initial screening by mail. Still, the Center accepted roughly 50 percent of submissions, and only 15 percent of clients reported a positive experience. After only one year, the Center was failing.

To turn the Innovation Center around, Oregon replaced Shaffer as director with Gerald G. Udell, an untenured professor with a degree in history and a professional background in marketing at General Electric. Udell brought a rhetorician's knack for selling the NSF's language of innovation. However, the former industry spokesman joined an academic department that looked at the Center with extreme suspicion.[35]

To bring order to the overwhelming number of inventor submissions, Udell developed the PIES (Preliminary Innovation Evaluation System) approach, a thirty-three-point checklist to help inventors, entrepreneurs, and innovators make "go–no go" decisions about further development of their ideas. The PIES innovation toolkit helped aspiring independents to understand five broad areas that contributed to the potential success or failure of their inventions: societal factors such as environmental impacts, business factors, demand analysis factors, market acceptance factors, and competitive factors. Udell argued that PIES could reduce the failure rate of new products by as much as 80 percent. Because his system put the onus on inventors themselves to make the "go–no go" call, face-to-face contact with clients became rare as submissions arrived from across the United States as well as Canada, Mexico, and Australia. Client satisfaction dramatically increased from 5 percent to over 95 percent, which proved that there was a robust demand for clear and concise innovation expertise.[36]

But improved client satisfaction did not correlate to successful commercialization. While the Center analyzed hundreds of ideas, only a tiny fraction proved viable. Few, if any, had a scientific basis. Among innovations that the Innovation Center touted as successes were a folding table that an inventor licensed to a furniture manufacturer, and the "Rare Ore" company, created by students who sold novelty pieces of low-grade gold.[37]

The Innovation Center also used its NSF support to establish entrepreneur education for University of Oregon students. The Center offered courses in Entrepreneurship, Management of Creativity, and Small Business Management and drew on students for assistance in evaluating submissions. Nearly 1,600 students enrolled in these courses, which Udell estimated would lead to 200 future entrepreneurs.[38]

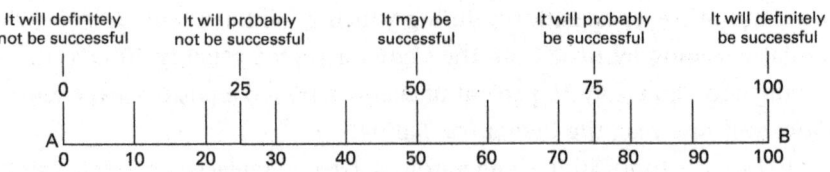

FIGURE 5.3

The Preliminary Innovation Evaluation System (PIES) used by the University of Oregon's Experimental Center for the Advancement of Innovation and Invention.

Despite evaluating over 4,000 ideas, the Center never developed a viable revenue model.[39] When the NSF funding ended, Oregon shut down the Experimental Center for the Advancement of Innovation and Invention. The experiment proved a high risk one for Udell, who was denied tenure. After a brief stint as an evangelical pastor, he resumed his career as an advocate for independent inventors. He created a private venture called the "Innovation Institute" that, in 1990, partnered with Wal-Mart to become the Wal-Mart Innovation Network (WIN). For a fee of $275, inventors would send a description of their idea and would receive a PIES evaluation.[40]

From the NSF's perspective, the Oregon experiment settled the role of the independents once and for all. The housewife or the laborer might aspire to invent, but they were not innovators.

CMU: FROM SCIENCE TO DESIGN

Pittsburgh's Carnegie Mellon University had all the components that experts believed were necessary for innovation: strong engineering, a world-class business school, novel approaches to urban planning with a focus on community needs, and a tradition of industrial design and the practical arts. CMU leaders claimed that they were smaller, younger, more interdisciplinary, and more agile than MIT.

The main reason that CMU received an NSF Innovation Center, however, was its director Dwight Baumann. Much as Everett Rogers had set the mold of the academic innovation expert and Hollomon the government innovator, Baumann exemplified a new kind of educator–entrepreneur. The son of North Dakota farmers, Baumann was another standout land-grant college student. After earning his engineering degree at North Dakota State University, he headed to MIT for his PhD in mechanical engineering where

FIGURE 5.4
Dwight Baumann. Courtesy of MIT Museum.

he discovered a love of teaching. He began working as an instructor in the undergraduate design division when he was just twenty-four. He also garnered a reputation as an unusually creative engineer with an interest in problems of the civilian economy and human needs. In the 1950s and 1960s, first as a student and then as an MIT professor, Baumann helped develop one of the first computer-aided design programs as well as computer interfaces for the blind and the nation's first "dial a ride" paratransit system.

Frustrated with MIT's culture of defense research and engineering science, Baumann left his tenured position in 1971 to join CMU to continue his work on transportation systems. There, he chartered the CED as an independent nonprofit organization. He then purchased the bankrupt Peoples Cab Company for a dollar, convinced CMU to use a former parking garage as an incubator space, worked across colleges to establish a master's program in engineering design, and drew up plans to use the cab company to educate designer–entrepreneurs.

Baumann's goal was to establish entrepreneurship education as a core function of a modern technological university. He was not especially concerned with making money and he cared little for nationalist rhetoric about global competition. Rather, his professed goal was to unlock human creativity in a "studio without walls, an association of people, loosely related, who communicate with each other and can get help when they need it." He bristled against the norms of engineering science and the notion of controlled experiments.[41] Technological innovation, he argued, could never be entirely predictable because it was a *project*, rather than an act of scientific discovery. "A project," he wrote, "is something that hasn't yet happened. And the instructors and students have the common goal of seeing how it'll turn out."[42]

The CED's mission was to advise and support entrepreneurs in the earliest stages of the innovation process when they needed space and seed funding. It created an environment for students to make a "sequence of nonfatal mistakes" so that they could fail and develop self-confidence for navigating the risks and uncertainties of entrepreneurial life. To shoulder the high financial risk of starting new ventures and to create formal barriers between the educational functions of the university, the CED operated as a private, nonprofit entity. Its approach was intentionally small, targeting graduate students who already had advanced scientific and engineering training and a viable idea for a business.

In CED's first five years, the center launched eleven ventures. In addition to the Peoples Cab Company, projects included a blood oximeter, a computer hardware company, and a newspaper printing technique. Many of these endeavors failed. Founders had personal health problems, patent disputes arose, and competitors claimed that CED gave an unfair advantage through the weight of CMU. CED distilled these lessons in brochures and public seminars, while faculty incorporated them into new classes.

A ten-point "readiness assessment" with over one hundred subquestions emphasized *personal* reflection before any technology or market evaluation. The first rule: "Only if you have sincerely made the decision within yourself to invest time and effort, and understand that sacrifice and risk are inevitable, should you consider the life of an entrepreneur." It aimed to show that innovation was a difficult path that could result in "personal dissatisfaction" and that one's "family goals" must not be sacrificed in single-minded pursuit of an entrepreneurial opportunity.[43]

CED also supported a few students who created successful start-ups. CED's breakout hit was CompuGuard, founded by electrical engineering PhD students Romesh Wadhwani and Krishnahadi Pribadi. The pair of international students from Pakistan and Indonesia spent eighteen months developing a security bracelet that used wireless signals to protect vulnerable people in dangerous work environments. But after failures to convert their prototype into a working design and severe cost overruns, they pivoted to a security and energy monitoring system for schools, prisons, and warehouses.[44] With CED assistance, the company secured government contracts and millions in private venture capital and grew to over one hundred employees. CompuGuard's first major client was the Los Angeles city schools. The two founders then sold the company in what at the time was the largest ever return on investment by a minority-run enterprise. Wadhwani became a serial entrepreneur, developing a successful robotics firm in the early 1980s. He is now one of Silicon Valley's wealthiest billionaire–philanthropists; his Wadhwani Foundation supports innovation and entrepreneurship education worldwide, particularly in emerging economies.[45]

When NSF funding ran out, a series of long-simmering tensions erupted at CMU. At the heart of most of the challenges was the cult of personality around Baumann, whose slapdash style conflicted with CMU's desire to compete with new technology entrepreneurship programs at the University of Pennsylvania's Wharton School and elsewhere. Baumann's onetime partner Jack Thorne took the lead of a new Enterprise Corporation, and Baumann was kicked out of his garage to make room for the initiative. Then, in the early 1980s, Baumann's department head rejected entrepreneurship as an "important area" for Mechanical Engineering. Baumann moved the CED to an abandoned YMCA building and attempted with limited results to help unemployed skilled laborers become innovators.[46] The CED faded, as CMU's faculty, like those at MIT, continued to fight over the proper role of university innovation and who had the authority to teach it.

UTAH: INNOVATING BEYOND THE COASTS

In testimony before the US House of Representatives, Wayne Brown, director of the Utah Innovation Center, identified Salt Lake City itself as one of the most important ventures in American history.[47] The Mormon engineer explained to the committee how founder Brigham Young had applied a recipe of education and self-sufficiency through which a few hundred pioneers had built one of the nation's great spiritual and economic engineers. The Utah Innovation Center, Brown argued, added a new ingredient to the original recipe, the industrial research park.

Utah's Innovation Center began in 1977 when the NSF already was evaluating data from its first three sites. Whether designed as an iteration on earlier trials, or an attempt to cook the results, Utah had the markings of a winner. The university already had an impressive track record for high-tech spinoffs. Artificial kidney pioneer Willem Kolff had joined the faculty in 1960, contributing to the region's moniker of "Bionic Alley." In 1965, former undergraduate David C. Evans returned to Utah to lead its new computer science department and, three years later, convinced Sketchpad inventor Ivan Sutherland to join him and launch a computer-aided design firm. They and their students would become central to the field of computer graphics over a span of decades, including Alan Kay (Xerox PARC), Ed Catmull (Pixar), and Jim Clark (Silicon Graphics, Netscape). Then, in 1968, the two professors launched the computer-aided design firm Evans and Sutherland Computer Company.[48] Utah's College of Engineering ranked a robust seventh in the nation in research dollars per faculty member.[49] Much of this activity was due to Brown, who had been Utah's dean of engineering prior to his role with the Center.

Utah's original Innovation Center plan focused on education rather than venture formation. Stephen Jacobsen, an inveterate tinkerer, had won the grant. Jacobsen's career demonstrates how small the world of university innovation was. He had completed his bachelor's and master's degrees under Brown and Kolff at Utah (although he was nearly expelled for a joke that almost blew up an engineering building). He then earned his PhD at

MIT for research on robotic control theory and body mechanics under the direction of Li and a colleague of Baumann.

However, the Center pivoted before the money even arrived. Jacobsen found himself overcommitted to other projects, so Brown stepped in.[50] An entrepreneur himself, Brown had founded several companies, including a firm that revegetated strip-mined extraction sites. Brown's strategy for the Center was to foster small companies in areas that matched the university's existing research strengths. It didn't hurt to have a former dean as director; under Brown's leadership, the Center hired a full-time administrator and moved from a small office to a shiny office park.

The Center had three objectives: it would create a consortium of key research and design labs on campus, it would develop a "technical assistance plan" for innovators and entrepreneurs, and it would run a joint education program through the Colleges of Engineering and Business. As part of its assistance plan, the Center offered offices, staff, legal aid, and capital. The Center was highly selective when choosing its tenants, selecting just eight out of one thousand proposals.[51] Students in the Innovation and Entrepreneurship course were partly responsible for determining product feasibility and designing plans for manufacturing, financing, and distribution. Center faculty and staff also founded the journal *Technovation* to spread their ideas about training innovators to readers in academia, industry, and government.[52]

The first companies incubated by the Center included Weathercaster, a graphics system for TV weather forecasting designed by meteorology graduate student Steven Root, and a model for an automatic, continuously variable speed transmission for bicycles developed by Robert N. Williams. Both, however, ultimately failed to thrive. The earliest true success was Bunnell Life Systems, Inc., which developed a life-saving ventilator for infants unable to breathe with their own capacity, and which remains a leading medical instrument manufacturer to this day.

When Utah's NSF funding ended, it faced a similar reckoning to its sibling programs at MIT, Oregon, and CMU. Where other NSF centers ended in failure or mixed results, however, Utah scaled its program to another level. Brown reinvented the Center as a private, for-profit company. As CEO and Chairman, Brown raised $1 million in capital from private investors to continue operations, and several clients, including Bunnell, moved with the Center into a massive 78,000 square foot incubation space. In its new incarnation, the Center focused on projects whose financial potential exceeded $50 million dollars. The Center also aided in the establishment of the Utah Innovation Foundation, an organizer of workshops and forums for Salt Lake City's business community.[53] Brown meanwhile became an evangelist for the NSF's Innovation Center Experiment. The Center created a satellite program in partnership with Oak Ridge Laboratories and Martin Marietta. Genexus, Inc., a new parent company, operated both the Oak Ridge facility and the Utah Innovation Center, and built a field office in Menlo Park, California.[54] The NSF's most successful test case thus ultimately found its innovation millionaires outside the university.

EVALUATING THE EXPERIMENT

As the Innovation Center Experiment wrapped up, the NSF patted itself on the back in a series of reports, conferences, and articles. "The ultimate effect of the Innovation Centers," it stated, would be "the regrowth of invention, innovation, and entrepreneurship in the American economic system."[55] Yet, by 1979, limited license returns from intellectual property had generated only $100,000.[56] Still, the NSF claimed that the experiment produced dozens of new ventures with $20 million in gross revenue, employed nearly eight hundred people, and yielded $4 million in tax revenue. Colton, the experiment's manager at NSF, imagined a national network of centers across fifty campuses producing hundreds of millions of dollars return on investment.[57]

The Innovation Center Experience garnered intense national and international interest from its inception to its completion. Established business schools in the United States created new technology-innovation tracks to compete with the experiments, while contingents from Canada, Sweden, and the UK traveled to learn about the model with hopes of recreating it.[58] On the other hand, Wisconsin Senator William Proxmire pointed to the banana peelers, video games, and sports equipment pursued in the Centers to lambast them as "wasteful federal spending" of "questionable benefit to the American taxpayer."[59] The *New Engineer*, a magazine for left-leaning technologists, similarly critiqued the program. It claimed that the biggest beneficiaries of the program were faculty rather than students, that private companies profited off students' work, that the government had funded new competitors to existing businesses, that many innovations were of "questionable technical merit," and that numerous conflicts of interest emerged within the centers.[60]

Finally, others criticized the narrow range of innovators in these hatcheries. At an NSF workshop, attended by congressmen, industrialists, and Center directors, African American chemist Grant Venerable lambasted the program for its narrow conception of innovation as the purview of white men at elite universities. "For me," he wrote, "it was a revealing experience to realize that apparently most technologists and mainstream innovators have not evolved to a state of consciousness which recognizes that women and excluded minorities are an integral part of the society in terms of career development and have particular needs and concerns." If supposed innovators could not address gender and racial equity "by more than a token nod," he wrote, "they are guilty of being part of the problem."[61]

External assessments of the Innovation Centers were mixed. Despite identifying numerous positive outcomes, one study concluded that the entire R&D Incentives Program was a failed experiment because political pressure and a lack of clear hypotheses resulted in "demonstrations" rather than true experiments.[62] In other words, the government's innovation programs focused on project-specific results rather than a systematic method that could be applied in future cases. The reviewers laid out seventeen criteria for rigorous experimentation and used the Innovation Center Experiment to illustrate their case. Colton defended the program arguing that, in fact, the centers met these criteria, but as "a socio-technical experiment" rather than a "strictly scientific experiment."[63] These claims highlighted the division between scientific norms and design norms. In the former, the gold standard was rigorously measured *validity*; in the latter, it was *viability*.[64]

Colton was partially warranted in his optimism. A second round of funding between 1977 to 1985 extended the University Innovation Center model to five additional universities. The expanded mission included innovation and entrepreneurship at a liberal arts college (UC Santa Cruz), turning research scientists into entrepreneurs (University of New Mexico), fostering small-town revitalization (University of Arkansas), rejuvenating traditional manufacturing (Worcester Polytechnic Institute), and developing appropriate technologies for rural communities (Oklahoma State). Responding to the pushback about the viability of these programs as scientific experiments, the NSF adopted the language of "demonstration experiments" and "design experiments."[65]

Although the centers represented a tiny fraction of the federal budget, they played an outsized role in national innovation policy. During the Carter administration, Democrat Adlai Stevenson III and Republican John Wydler cosponsored a bill creating Centers for Industrial Technology at universities and nonprofit organizations.[66] In congressional testimony, the University Innovation Center directors shared their lessons learned. Baumann insisted that technology transfer was, above all else, a problem of human assistance, while Udell warned to expect resistance from those who believed that university innovation centers were "the antithesis of the true function of the university."[67] Carter signed the Act into law in the late stages of the 1980 election; however, after his defeat, the proposed Centers were not created.

University incubators proliferated during and after the center experiment. But had they worked? In 1985, the NSF contracted a final study by a private research company that interviewed, surveyed, and compared the innovative activity of nearly six hundred alumni; half who had been trained in the centers, and half who had not. The study identified common problems across campuses: the lack of faculty rewards; a "perceived illegitimacy of assistance to profit making;" and bureaucracy, leadership, financial instability, and turf wars. It found that while most Centers went through periods of uncertainty and drift, all but Oregon continued to operate in some capacity. Surveys found that the Centers provided a high level of service to clients and contributed to venture success. The biggest unknown, the study argued, was whether the "traditional split" between "academic values and entrepreneurial values" could be bridged. Here, the report let the numbers do the talking: 31 percent of students with Innovation Center experience subsequently created their own invention and 35 percent started their own business, a nearly 50 percent increase among students who had spent time in the Centers versus the control group who had not.[68]

WHAT HATCHED

Incubators are nurturing devices that create safe environments at a vital life stage. They are monitored with care by knowledgeable experts. The NSF Centers applied the tool to human potential. Where the government previously used mechanistic and militaristic images of pipelines and manpower, the Innovation Center Experiment spoke of the "nourishment" of "fledgling" students in an environment that stressed empathy over evaluation. The experimenters encouraged students so that they could try, fail, and find their "passion." Young innovators were a rare breed to be cultivated, not crushed in the nest.

But incubators are not built to serve the chicken, they are tools for agricultural yield. Incubators create an artificial environment that mimics nature to prepare lifeforms for broader utility. They mitigate risk, but place bets by design. Those in the University Innovation Center Experiment with the most experience in the entrepreneurial economy spoke of "overwhelming odds" and stressed the importance of helping would-be innovators to recognize when they lacked what it took whether in character, idea, or conditions to succeed. The higher up the university and government org chart one traveled, the more intellectual property and "exploitation" characterized innovators as a new form of scientific manpower; not raw material, but speculative capital. This, after all, was built into the NSF's experiment diagram.

The Center hatchlings died at a high rate. Oregon's Center could not buffer independent inventors from the risk, so it helped their dwindling lot to predict their fate. The simulated environment of MIT's Co-op could only do so much to prepare bright-eyed students for the reality of bankrupt partners. Moreover, some of the NSF's "embryo enterprise" centers like Oregon and MIT themselves failed to thrive.[69]

Still, it is not difficult to see what attracted faculty and students to the centers. Participants generally *were* the venturesome misfits that policymakers sought to elevate to national prominence. They came to the experiment out of desires to reject the military–industrial complex, to find solutions to social problems, to design and build things, to make a buck, and to pursue science and technology as a personal vocation. They built energy-efficient insulation and solar panels, games and toys, medical devices, and urban transportation systems instead of aircrafts and bombs.

The legacies of the University Innovation Center Experiment are visible on nearly every college campus today. In 2011, for example, the NSF dramatically expanded the approach when it created the Innovation Corps (I-Corps), a "hypothesis-driven business-model discovery" method for training university innovator–entrepreneurs. The program, which puts a faculty member, a student, and an industry partner through a six-week "boot camp" has spread to dozens of campuses and trained thousands of teams with many of the concepts developed in the Innovation Center Experiment, including, for example, Oregon's emphasis on "go–no-go" decisions.[70]

What also hatched was a robust conception of the entrepreneurial self. Innovation experts' models and desires altered not only the function of the university, but also "the ego of the person."[71] They institutionalized powerful images of the scientific innovator–entrepreneur as a risk-taker who understood the probabilities of capital just as well as thermodynamics. The ideal innovator–entrepreneur sometimes used institutions as a launching pad for private gain, but in service of the aggregate good. These values are now so codified that a recent textbook on technology entrepreneurship uses them to place the entrepreneurs as "agents of progress" at the center of all economic growth.[72] The purpose of innovation education, however, was not just about breeding winners. All students, even those who never intended to commercialize their ideas or build a start-up company, would benefit from learning to be *entrepreneurial* rather than *entrepreneurs*. "Career resilience and fulfillment," my own university announces, are a twenty-first century requirement because "students will need to adapt as our economy evolves and new jobs and careers are defined."[73]

These impacts of the Innovation Center Experiment were not obvious when the program closed its first round in 1978. Innovation education had its evangelists and its misfit practitioners who taught others in their ways. These educators found ambivalent support at the federal level as presidential administrations from both parties experimented at a small scale. University faculty and administrators in the 1970s were more apt to view such programs as frivolous, nonacademic, or not worth the investment than they were to see the hatcheries as a threat to the moral purpose of the university or conversely as the fuel of its economic engine. All of that was about to change—not due to the success of the hatchlings, but rather out of fear that the entire national yield was at risk.

II   DEFICITS AND DREAMS

6 CREATIVE CLASS CONSCIOUSNESS

The erosion of our confidence in the future is threatening to destroy the social and the political fabric of America.
—President Jimmy Carter, 1979

It's time for the Creative Class to grow up and take responsibility. But first we must understand who we are.
—Richard Florida, 2002

In 1989, the Massachusetts Institute of Technology assembled its collective brainpower to make a "dynamite" announcement. For the first time since World War II, its faculty banded together in the nation's service. Rather than unveil a breakthrough invention or world-changing discovery, however, they warned of a dire threat. According to the MIT Commission on Industrial Productivity, after two decades of crises and missed opportunities, the United States had fallen dangerously behind in the global economy. Its five hundred–page exposé, *Made in America: Regaining the Productive Edge*, described broken government, parochial managers, and disgruntled workers amid rapid technological change. Led by computer scientist Michael Dertouzos, nuclear engineer Richard K. Lester, and economist Robert Solow, a team of experts spent three years interviewing hundreds of stakeholders on multiple continents. The headline conclusion: "We have no one to blame but ourselves."[1]

According to journalists at the time and to scholars since, the 1980s were a watershed for American innovation. Silicon Valley became a leading force in the global economy, making celebrities out of its geek founders as Americans introduced computers into their homes. University commercialization of science flourished in the wake of the Bayh–Dole Act and its granting of patent rights to federally funded discoveries. The biotechnology start-up Genentech received FDA approval for its life-saving synthetic insulin. MIT itself contributed to and reaped the benefits of this innovation boom. Just months prior to *Made in America*'s release, cultural tastemaker Stewart Brand had lauded the da Vinci–like innovators at MIT's new Media Lab for "inventing the future" in a novel fusion of art and technology.[2]

Why did MIT's innovation experts see looming disaster where others foretold digital utopia? The crux of the MIT Commission's warning was a technology-driven transformation in the global economy. The Commission explained how Japan and Germany, once decimated by World War II, had made rapid gains in the 1970s via new approaches to manufacturing. Meanwhile, international trade was accelerating due to a combination of national deregulation and advances in communication and transportation. The catch-up was an indisputable good, but its benefits were unequally distributed, creating a gulf between globalization's winners and losers. The United States increasingly was on the wrong side of the gulf. The consequences were most devastating in manufacturing where joblessness plagued once thriving regions. *Made in America*'s experts warned that the same pattern that had led to a decline in steel and automobiles was now threatening the microelectronics sector. Higher education was part of the problem. With a few exceptions, universities had abandoned practical technology for ivory tower science. The result: other countries innovated while the United States suffered from stagnation, bloat, and indecision.

Made in America garnered national headlines because of MIT's reputation, but also because it confirmed what pundits and policymakers already took as truth. Throughout the 1980s, a chorus of innovation experts, politicians, and journalists asserted that the entrepreneur's mantra "Change or Die!" applied to the United States itself.

This chapter explores how once experimental aspirations for innovation policy became a national—and nationalist—imperative. Central to the transformation was an emerging consensus that the United States was in decline because its innovation engine had slowed. The chapter begins in the 1970s when experts in the Ford and Carter administrations revisited President Nixon's interrupted innovation programs. It then shows how, in the 1980s, these experts became central players in American policy by using the competitiveness crisis to scale up initiatives that they had piloted in the two decades prior. The defining terms of their vision were *competitiveness* and *productivity*, and their central message was of a *deficit* between the United States and other industrial societies.[3]

The chapter then demonstrates how the hard-headed industrial policy by engineers and economists evolved into a cosmopolitan vision that linked innovation to liberal values of tolerance and diversity. At first, innovation policy focused on infrastructure rather than the exploits of individual innovators and entrepreneurs. But as innovation policy evolved, a different group of experts claimed that deficiencies in technology and economics were rooted in a deeper deficit of culture. By the late 1990s, experts argued that the dynamics of communities that attracted supported, and multiplied talented innovators were essential to regional success and national renewal. Led by geographer Richard Florida, they argued that creatives at the core of innovative communities constituted a new social class responsible for a transformation as great as the industrialization that led Karl Marx to predict an inevitable revolution.

INNOVATION POLICY COMES OF AGE

In the late 1970s, innovation experts in and around the federal government built on Kennedy and Johnson era foundations, but with little in common with that era's liberal optimism. Where earlier federal champions viewed innovation as the source of a dynamic civil society, policy experts in the post-Nixon era focused almost entirely on technology and economic growth as a hard-headed response to a grim new reality.[4] These experts' pessimism was grounded in their experience of the first presidential resignation in American history, an energy crisis that revealed dependence on foreign oil, the shuttering of space exploration, job losses due to automation and offshoring, and stretches of double-digit inflation.

These industrial innovation advocates were corporate executives and science administrators concerned with restoring investment in science and technology to Cold War heights. Innovation policy's inner circle included systems engineer and aerospace executive Simon Ramo, Texas Instruments

CEO Patrick Haggerty, former Nixon advisor and Exxon Mobil senior vice president and Edward E. David, and IBM Chief Scientist Lewis Branscomb, a physicist who had a long career at the Department of Commerce before decamping to industry. These technocrats targeted the civilian economy but represented the military–industrial complex's appropriation of a rhetoric once formulated against it. The crux of their argument was that the United States needed to coordinate its R&D and manufacturing through industrial policy.

The White House's Office of Science and Technology Policy (OSTP) served as an incubator for the innovation policy experts. In 1975, President Ford restored the science policy unit which Nixon had abolished. Among OSTP's first initiatives was an advisory group on Science–Technology–Economic Policy Interactions, or "technology–economics" for short. Led by Ramo, the technology–economics group set out to demonstrate that "international economic competitiveness" derived from science and technology and that flatlining federal investment had made the nation's science "less bold, less innovative, more timid."[5]

Under President Jimmy Carter, the technology–economics group expanded its inquiry. The group brought Hollomon back into the fold when it contracted his MIT think tank, the Center for Policy Alternatives, to produce the world's first comparative study of national innovation regimes. Hollomon assembled a team of international scholars who analyzed the US, France, UK, West Germany, and Japan. The report revealed tepid outcomes but sounded no alarms. They concluded that perhaps apart from Japan, innovation policy worldwide was "ad hoc" and "palliative," and repeated the message from the 1960s that innovation was important, but no nation knew how to plan for it.[6]

THE INNOVATION DEFICIT

Amid the economic stagnation of the late 1970s, the technology–economics experts honed ambiguous research about innovation policy into an alarming message of an international deficit. In Congressional testimony, Phil Handler, president of the National Academy of Sciences boasted of past achievements. "Perhaps 80 percent of all of science has been learned since the birth of the National Science Foundation," he declared, "while other peoples were still living in the Bronze Age."[7] But, according to Handler, the

United States's supremacy was over. A diminished science-policy apparatus, parochial manufacturing industries, high labor costs, and esoteric university research were symptoms of an outdated innovation infrastructure.

Journalists amplified the message, decrying "Vanishing Innovation," and the "Sad State of Innovation." In classic Washington fashion, these articles were veiled proxies for legislative aims; they were filled with quotes from the technology–economics experts and introduced into the Congressional Record by the politicians who supported those technocrats.[8] *Newsweek*'s "Innovation: Has America Lost Its Edge," is particularly instructive. The cover story opened with a warning from Senator Stevenson that the American "spirit of adventure and invention may be drying up," adding that "nations fail when that happens." In a style reminiscent of *International Science and Technology*, a comic strip took readers on a journey through the red tape of "taking a product from lab to market." In this morality tale of frustrated dreams, scientist I. M. Genius discovered an energy-saving formula that promised to dramatically boost gas mileage thus saving money and the environment. However, Genius's company Chemtrust learned that a Japanese firm was developing a similar innovation. So Chemtrust filed a patent and consulted with federal regulatory agencies. Unfortunately, Ralph Riskless at the EPA demanded additional testing that added two years and $2 million to a project already over budget due to rising inflation. Meanwhile Chemtrust invested $20 million in a new plant, but resistance and litigation added another $10 million. Construction workers then went on strike. Finally, after almost ten years and $50 million, the energy-efficient product reached the market . . . but so did the Japanese competitor's—at half the price because it was aided by proactive government coordination.[9]

Federal hearings, think tanks, and political pundits affixed blame for the deficit and offered prescriptions to resolve it. In 1978, the Carter administration called for enhanced *industrial innovation* in a bipartisan report authored by 150 leaders across government, industry, academia, and labor. However, tensions simmered beneath the surface as innovation boosterism competed with environmental protection and labor rights. Buried in the report was a dissenting public interest subcommittee analysis which concluded that regulation spurred, rather than deterred, innovation. It concluded that the government needed "public social accounting" to ensure that the benefits and tradeoffs of innovation were equitably distributed.[10]

FIGURE 6.1

A cartoon guide to the innovation deficit from *Newsweek*'s "Innovation: Has America Lost Its Edge." The original two-page layout here combined into a single image. Illustrated by Roy Doty.

With the 1980 presidential election looming, technology–economics experts used the scourge of a deficit to achieve coordinated reforms. Conservative engineers shared common cause with "neoliberal" intellectuals, such as MIT's Lester Thurow, and "Atari democrats" who championed information technology as the nation's economic future.[11] With both Democrats and Republicans campaigning as the innovation party, Congress passed a

suite of legislation including the Stevenson–Wydler Technology Innovation Act, to foster technology transfer from government laboratories, and
the Bayh–Dole Patent and Trademark Law Amendment Act, aimed at com-
mercializing discoveries funded by government research through the granting of patent rights.

Ramo and the technocrats hoped to go further. They wanted to create a
National Technology Foundation (NTF) as a parallel agency to the NSF that
would integrate policy, business development, and the Patent Office under
one roof. The engineers vowed to expand innovation programs that the
science-dominant NSF treated like "stepchildren."[12] The NTF effort failed,
but as a consolation prize, the NSF established an engineering directorate.
And, in an act of double reconciliation, Carter appointed John Slaughter as
the NSF's first Black director as well as its first engineer.[13]

Ronald Reagan's victory empowered the technology–economics experts
who were making innovation synonymous with a planned economy. Even
as Reagan advocated for reduced government, his administration expanded
its hand in industrial policy. Serving on the president's Task Force on Science and Technology, Ramo and David aimed to replace the basic science
model with an "Edisonian idea of purposeful science."[14]

In 1985, the Reagan administration identified the innovation deficit
as a primary cause of a global competitiveness crisis. A blue-ribbon Commission on Industrial Competitiveness, led by Hewlett Packard president
John A. Young, an electrical engineer who had been with the Silicon Valley juggernaut since the 1950s, called for a response that rivaled the space
race. The Commission's report, *Global Competition: The New Reality,* identified four core needs in the American economy: technology development,
capital flow, a modernized workforce, and international trade. They argued
that the government's role in reversing the innovation deficit was as a
"consensus-building" nurturer and as a maintainer of "steady" rather than
disruptive growth.[15]

Global Competition put a new shine on an old idea about international
trade. In the nineteenth century, British economist David Ricardo argued
that when countries specialized in certain industries—English cloth and
Portuguese wine, for example—instead of individually trying to produce
both simultaneously, the total production increased and both nations benefited.[16] Updated for a technological age, this meant that America's competitiveness crisis was not "a winner-take-all game," but rather one in which all
nations could thrive by maximizing their "comparative advantages." The

United States' biggest advantages were in technology and its skilled work-force.[17] But the advantages were relative and ever changing. Moreover, the country's high standard of living paradoxically was a disadvantage because other nations could pay lower wages.

For the technology–economics experts, *productivity* was the most important variable for maintaining the nation's advantages. Since the 1920s, productivity—defined as the ratio of output per unit of input in the economy—had been used to study the relationship between labor and growth. However, in the 1950s, Solow, Kenneth Arrow, and other econ-omists shifted the meaning of productivity from workers to technical change. Innovation policy experts in the 1980s breathed sociological life into Solow's abstract production function. *Global Competition*, for example, reported on the attitudes of workers, claiming that Japanese employees were ten times more likely than their American counterparts to believe that they would benefit personally from their companies' productivity gains.[18]

Productivity was no mere buzzword; it was a scientific object that could be measured, benchmarked, and evaluated. *Science* magazine became a leading innovation policy venue. In its pages, for example, wunderkind economists Paul Krugman and Lawrence Summers teamed up with engineer–entrepreneur George Hatsopoulos to decry a drop in annual labor productivity growth from 1.9 percent in the 1960s to 0.3 percent after 1973.[19] Japan was again a threat and a model. Hiroshima University economist Shigeo Minabe explained to *Science*'s mostly American readers that Japan's government steered its indus-tries as a "player-coach" rather than an "umpire," resulting in less waste, less stagnation, and greater creativity.[20]

Innovation policy advocates went out of their way to claim that their reforms would benefit American workers, even as they argued that the pres-sures of technology and international trade demanded rethinking labor relations. Amid retrenchment in union power, experts glossed over labor conflicts and promoted cooperation. A 1987 National Academies report, *Technology and Employment: Innovation and Growth in the US Economy*, for example, contended that productivity was the key to good American jobs. Headed by Carnegie Mellon University president Richard Cyert and econo-mist David Mowery, the effort sought consensus among stakeholders from the Department of Labor, the American Federation of Labor, major cor-porations, and private foundations. The coalition conceded that creative destruction generated inevitable job losses and inevitably fostered labor resistance, protectionism, and wage stagnation. But, they argued, the failure to close the innovation deficit would have severe consequences for American workers. If the US could maintain its innovation advantage, it could generate higher productivity and high-wage jobs. They advocated for more federal support, continuing education, and cooperation to offset job losses due to obsolescence. "In the end," the National Academies sanguinely concluded, "no trade-off need be made between the goals of high levels of employment and rapid technological change."[21]

By the late 1980s, critiques of the innovation policy consensus were few and far between. Politicians and experts debated the specific means of improving productivity, but they were unanimous about innovation's role in national prosperity. During the 1988 presidential election, for example, George H. W. Bush declared that "Technology is America's economic fountain of youth," while Michael Dukakis made the "Massachusetts Miracle" of high-tech revitalization his core campaign pitch.[22] Military–industrial technocrats like Ramo and David elevated engineers into science policy roles once held by physicists. Economists such as Summers integrated tenets of an innovation deficit into neoliberal economics. Innovation policy was embraced equally by liberal institutionalists such as Mowery and Richard R. Nelson who rejected the "simple minded arguments" of "zealots." These more cautious experts argued that any one nation's success was not easily copied, and that high-tech policy programs often failed. They nonetheless reached the same conclusion as the zealots: the United States needed a national innovation system.[23]

REPAIRING THE DEFICIT

With the deficit identified, government, industry, public–private coalitions, and universities set upon a campaign of national maintenance and repair through innovation. The competitiveness crisis proved a unifying call because it translated across all strata of society. Success at the top depended on change at the bottom, and vice versa—there were deficits all the way down.

MIT's *Made in America* codified the scalar quality of the deficit and the strategies for closing the gap. Its technology–economics experts argued that fixing the nation's problems required understanding the "ground truth" of manufacturing. Abstractions of economics revealed themselves to be practical problems "on the shop floor, in the laboratory, in the boardroom, and in the classroom." Surveying eight industrial sectors, MIT's team found deficiencies everywhere.[24] Unfortunately, they concluded that there were no simple solutions. Relying on a service economy to enhance productivity would not be sufficient, nor would offshoring or outsourcing. To remain competitive, the United States would have to innovate everywhere at once.

The federal government pursued three kinds of reforms to foster Edisonian science.[25] First, it implemented regulatory changes such as Bayh–Dole and the Orphan Drug Act to enhance the climate for enterprise. Second, it made widespread reforms to the federal R&D network that included the public–private semiconductor manufacturing initiative SEMATECH, which sought to emulate Japan's industrial policy. Third, it enhanced its investment in entrepreneurship incubators, notably through the creation of the Small Business Innovation Research (SBIR) program.

The SBIR program illustrated the complex goals of the emerging national innovation system. SBIR was created to resolve the twin deficits of technological innovation and well-paying jobs by acting as a public venture capitalist. It aimed to generate breakthrough technologies by making thousands of tiny federal investments in fledgling companies.[26] SBIR required any agency with an innovation component—including DOD, DOE, EPA, and NSF—to direct 1.25 percent of its R&D budget to businesses with fewer than five hundred employees. In the 1980s, SBIR granted over 14,000 awards for $2 billion.[27] But how would these start-ups generate jobs? The employment rationale for SBIR came from a Department of Commerce commissioned study by MIT innovation expert David Birch, which upended the conventional wisdom that large corporations were the major producers of new jobs. Utilizing a massive data set, Birch instead claimed that small businesses had a disproportionate impact on job growth and regional stability.[28]

In tandem with these federal programs, nearly every state in the union adopted a variation on the dream of economic renewal through technology. Although federal funding served a coordinating role, state and regional innovation programs were not merely a trickle-down diffusion.[29] Some states had home-grown innovation programs dating to the 1960s. Others, especially in Southern states hostile to federal initiatives, established new local programs. Outside coastal technology hubs, states banded together in regional coalitions such as the Southern Technology Council and the Council of Great Lakes Governors. With names like Ohio's Thomas Edison

Program, these programs promoted a manufacturing renaissance based on local dynamism. In the 1988 fiscal year alone, such programs distributed $550 million in public investment.

Pennsylvania's Ben Franklin Partnership, one of the most heralded coalitions, showcased the goals and programs of regional innovation initiatives. Formed in 1982, the program merged technology-economics policy with local extension programs to foster "job creation, job retention, and regional economic growth through the commercialization of innovative technologies."[30] Modest grants targeted small business creation in advanced materials, manufacturing, and environmental technologies. Businesses would keep the intellectual property and pay universities with royalties. Because job creation was the goal, grant recipients were required to locate manufacturing in the state.

Universities played a starring role in regional initiatives that built on their experiments in the 1970s. Administrators and faculty positioned their institutions as the ultimate advantage makers; not just as sites of discovery, but also as proffers of innovation expertise, collaboration brokers, and trainers of innovation millionaires.[31] Where earlier university programs targeted small pockets of students and faculty, initiatives in the 1980s sought to transform departments, colleges, industries, cities, and state economies.[32]

In 1985, to spur local efforts through federal planning, the NSF launched the Engineering Research Center (ERC) program for "cross-disciplinary, systems-oriented engineering research on problems critical to industry."[33] ERC supported manufacturing-oriented research centers for up to $25 million at a single school, investing in robotics, composite manufacturing, telecommunications, biotechnology, intelligent manufacturing, and systems research. Meanwhile, faculty witnessed colleagues like Genentech's biochemist founder Herbert Boyer become celebrity millionaires. Journalists also picked up on the trend, declaring that the campus laboratory had replaced the neighborhood garage as the leading incubator for start-ups.[34]

A CULTURE DEFICIT

Initiatives spawned by industrial policy helped produce a new cohort of innovation experts who developed an important twist on the deficit model. In contrast to the military–industrial backgrounds of the technology–economics group, these specialists came from interdisciplinary fields such as geography, urban studies, and organizational planning. This cohort included the first cluster of women innovation experts: MIT political scientist Susanne Berger, who helped coordinate and ghost-write *Made in America*; Marta V. Goldsmith of the Urban Land Institute; AnnaLee Saxenian of Berkeley's Department of City and Regional Planning, a former student of Berger's; Saxenian's prolific Berkeley colleague Ann Markusen; and Maryann Feldman, author of the *Geography of Innovation*.[35]

Regional innovation experts complicated the technology–economics model by arguing that productivity depended as much on cultural dynamics as it did money and technology.[36] These experts extended the concept of comparative advantage to its next logical step: just as nations struggled with ever-shifting inequalities, manufacturing cities declined while high-tech clusters boomed.

Saxenian's *Regional Advantage: Culture and Competition in Silicon Valley and Route 128* was the genre's breakthrough hit. Combining anthropology with business advice, she explained how the competitiveness crisis challenged the United States' top technology regions but also why outcomes differed between them. Where Silicon Valley continued to thrive, Boston's Route 128 was in decline. Why? Saxenian's answer was *culture*. Economies were shaped by their social and institutional dynamics, "the shared understandings and practices that unify a community and define everything from labor market behavior to attitudes toward risk-taking." These dynamics differed from region to region. Silicon Valley's organizational culture was "laid back" and valued "collective learning and flexible adjustment." Rather than limit themselves to individual companies, Silicon Valley's innovators had built a network of "collective technological advance." Route 128, in contrast, reflected Puritan values of self-sufficiency which resulted in distrustful, independent firms. Saxenian dubbed this difference "collaborative advantage."[37]

Like cloth and wine, however, innovative regions succeeded through differentiation. That meant that Silicon Valley could not simply be copied and that established centers like Route 128 could falter. The flipside was that every place had a unique set of ingredients. Saxenian personified regions to argue that innovation gains required a difficult process of "self-understanding" in which communities cultivated their best selves.[38]

The innovation deficit was a culture deficit.

MANUFACTURING CULTURE

Immersed in blue-ribbon reports, state hearings, and economic indicators, what happened to the innovators? The technocratic bent of innovation policy had rendered *Newsweek*'s "I. M. Genius" into a cog in the industrial system. Government planning emphasized incubating new businesses but focused on aggregate data of return on investment, companies assisted, and jobs created. Innovation policy experts nonetheless had a transformative impact on the imagination and lived experience of innovators.

Both the Carter and Reagan administrations rekindled Nixon's efforts to use national prizes to promote innovation's heroes. President Carter authorized a President's Innovation Award that under President Reagan became the National Medal of Technology. In contrast to Nixon's intended awardees, its first winners all came from industrial corporations, including Boeing, IBM, and Philips Petroleum; and most of the winners worked in teams rather than as solo innovators.[39]

The growth of university innovation programs also had a far-reaching impact on the identity of their practitioners. Technocratic efforts like the NSF's ERC program were also culture change projects that institutionalized new ways of being scientists.[40] Running throughout these initiatives was a faith that university-trained innovators could connect East Coast manufacturing with West Coast information technology, with the concept of "multidisciplinary design" reconciling these worlds. Programs at MIT, Stanford, and CMU each developed major new programs to capitalize on the agglomeration of talent in innovative regions.

To understand how the deficit model became entwined with visions of creative renewal on American campuses in the 1980s, it is helpful to explore the era's most celebrated incubators of innovation culture.

MIT: LABORATORY FOR LEONARDOS

MIT, as we have already seen, had a complicated relationship with innovation culture. In many respects, the technological institute was the epicenter of American innovation. Since the 1930s, MIT had an industrial liaison program to match local companies with academic researchers. In the 1940s, it helped to create the venture capital firm ARD. MIT was also home to cross-disciplinary programs in economics, architecture, urban planning, management, and the arts and humanities. These programs housed prominent innovation experts, including Solow, Hollomon, Schön, Bennis, and Baumann to name just a few. And yet, MIT was a scientific powerhouse. In the twentieth century, federal contracts and Nobel Prize–winning professors dominated MIT's institutional culture, prompting faculty in otherwise elite programs to lament their subordinate status and overlooked contributions.

The competitiveness crisis of the 1980s offered a reversal of fortune for MIT's innovation experts. MIT's Commission on Industrial Productivity accused its own institution of contributing to the innovation deficit. MIT's best students, its authors wrote, "were not drawn into manufacturing, let alone . . . the mature, conservative, grubby world of machine builders." Repairing the situation required a shift from ivory tower eggheads to practical engineers trained in multidisciplinary design.[41]

The reversal, however, was not the grubby one imagined by the technology–economics experts. Instead, in 1985, a new campus center, the MIT Media Lab, offered a vision of design as a cosmopolitan merger of science, engineering, and the creative arts. Founder and director Nicholas Negroponte positioned the Media Lab as a home for boundaryless exploration where distinctions between scientist, artist, engineer, architect, and philosopher no longer applied. Negroponte asserted that issues of economic competitiveness were in fact issues of merging manufacturing with the digital economy, which were problems of new collaborations, which were problems of technology and creativity, which were problems of human desires and needs, which were problems of societal good.

The Media Lab blended humanistic and technocratic themes of innovation expertise in a convergence of history and futurism. In the two decades prior to the lab's founding, MIT had invested in the creative arts to humanize its scientists and engineers and to find affinities between scientific discovery and aesthetic practice.[42] Architects, social scientists, and engineers at MIT had also introduced computational techniques into their work as computers shifted from mathematical calculation to graphics and sound.[43] Finally, MIT president Jerome Wiesner supported Negroponte's vision as a path to MIT's post–Cold War future.

The Media Lab was built on a new model of corporate sponsorship. It had over one hundred partners that included Digital Equipment

Corporation, the nation's major television networks, Apple, DARPA, IBM, and LEGO. Negroponte boasted that rather than secret or proprietary deals, these sponsors gained access to the entire lab. The Media Lab emphasized cutting-edge demonstrations of near-future breakthroughs. These included projects aimed at industrial production such as using holograms to design cars. But the Media Lab stressed human–machine collaboration for people of all walks of life from electronic music for composers to programming languages for children.

The Media Lab was a press darling for its futuristic optimism. It didn't hurt that most major publishing, broadcast, and entertainment corporations were clients. But the Lab's visibility stemmed from its human-centered vision. A fawning book by Stewart Brand concluded that the Lab represented the best hopes for a digitally mediated culture, repeating Negroponte's claim that the Lab was dedicated to "quality of life in an electronic age" and that it offered "a primer for a new life-style."[44]

Brand's book reinforced the image of the ideal environment for innovation as one that supported collaborative collisions among transgressive researchers. Its innovators worked with a "Demo or Die!" ethos in an organization without silos. They were as much at home in a corporate pitch setting as they were in an art collective. Apart from a few senior luminaries, Lab members were under forty years old. Negroponte was at turns "an exotic with the moves of a jet-set executive and a businessman's get-on-with-it rigor." Hologram expert Steve Benton "look[ed] like he strayed into his office from a beer commercial—a big, hearty, athletic-looking, humorous guy." Composer Tod Machover was a "cheery young composer . . . of exceptional talent." Christopher Schmandt was a "former college drop-out who spent five years hitchhiking across Africa, the Mideast, and India." Brand put a name to the colorful residents of this ecosystem; the innovator of the information age was an "amphibian."[45]

STANFORD: ANYONE CAN DESIGN

At Stanford, a design-based innovation culture emerged under similar conditions to those at MIT. In the 1990s, a group of educators elevated a once marginal activity to a key feature of Stanford's identity. Where MIT emphasized exclusivity, however, Stanford's innovation experts sought to democratize participation by insisting that anyone could design.

FIGURE 6.2
The MIT Media Lab as a home for innovators. Top: Nicholas Negroponte and Stephen Benton. Center: Tod Machover and Seymour Papert. Bottom: Walter Bender and Alan Kay. Photos on laserdisc by Marie Cosindas.

Design at Stanford got its start in 1958 when the university poached mechanical engineer John Arnold from MIT. Arnold borrowed techniques from psychology and business to present students with outlandish challenges that included designing automobiles for aliens. At Stanford, he cultivated a small faculty that included mechanical engineer Robert McKim, an advocate of using LSD to stimulate creative problem-solving; Matt Kahn, a twenty-something art professor; and Bernie Roth, an engineer radicalized by the antiwar and appropriate technology movement. Just as the program was getting started, however, Arnold died of a heart attack.[46]

James (Jim) L. Adams, one of the program's first graduates took over for Arnold and helped to turn Stanford's niche program into a national leader. Adams was an amphibian by training. His career began conventionally with an engineering science degree from Caltech. After college, however, Adams left engineering for art school. He then abandoned that path to join the Air Force. After his service, Adams enrolled in Stanford's new program with the hope of merging his technical and artistic interests. Adams earned his PhD in 1961 and landed a prestigious job on the design team for the Apollo spacecraft. However, his passion for teaching drew him back to Stanford, where he would work for nearly half a century. Adams became a beloved teacher and administrator who would chair multiple departments and programs.

Adams believed that the purpose of design was to make technology serve human values. In the late 1960s, he led a social-justice-oriented undergraduate program to educate students to "solve poorly defined, open-ended, multivariable and cross-disciplinary problems."[47] He argued that doing so required nurturing multidisciplinary graduates who could combine "the needs and abilities of people" with technical tools. At the height of campus revolt, Adams argued that universities were moving from a "thing" orientation to a "technology and society" orientation. This shift would be accompanied by a demand for "lawyers with an appreciation of technology and the ability to dance" and "writers with an appreciation of biology who can get into Congress."[48]

The Design Division's philosophy was in opposition to the dominant culture of Stanford's engineering science or economics faculty who view design as suited to "second and third echelon schools."[49] In a pair of unusual engineering textbooks, Adams and his colleague McKim argued that engaging multiple senses was vital for approaching problems in new ways, not taking givens for granted, and not rushing to solutions.[50] This "alternative thinking language" had been lost amid the valorization of mathematical rigor and computerization. Design, they argued, was a philosophy and a method for "thinking well."

During the competitiveness crisis of the late 1970s and 1980s, the Design Division extended its ambitions for thinking well from specialist students to industry professionals. In 1978, the Division created an executive short course on design methods for technology managers.[51] It also found allies and collaborators who shared its holistic approach to design amid broader growth in industrial design and human–computer interaction in Silicon Valley, especially at nearby Xerox PARC.[52] One of the Division's dropouts greatly enhanced its reputation and expansion as a hub in Silicon Valley's design culture. In 1975, David Kelley, a wayward Boeing engineer, joined the PhD program. Kelley thrived as a problem-solver and teacher in the Design Division but struggled with academic research. In 1978, he abandoned his dissertation on medical information systems to start the consulting firm that became IDEO, one of the most recognizable design companies in the world. But Kelley never entirely left Stanford. As his consulting business boomed, he cotaught courses such as a "design for extreme affordability" with his former teacher Roth.[53] A campus fixture and a friend of senior administrators, Kelley received tenure in 1990, despite not completing his PhD and never publishing a paper.

By the early 1990s, the Design Division was a national leader that had produced two hundred alumni in fields as varied as product design, medical rehabilitation, and industrial robotics. Looking for new ways to grow, the Division offered its first Design EXPerience in 1993. The managerial short course brought academics, executives, and journalists together for lectures and team-based improvisational theater akin to those of the Innovation Group before it. The event lost $20,000 but was a catalyst for a new brand of innovation expertise. Subsequent iterations included an affiliates program that showcased students' designs on projects for corporate partners and served as a de facto hiring program.[54] By the end of the decade, promotion for the Design EXPerience echoed the promises of the innovation geographers:

> Call it geography. Climate. Culture. It's a mélange of intellectual stimulation, history, luck, friendships, willingness to risk, people-sharing and idea-swapping. It's a sense of community and a habit of thought. Increasingly, it's global in perspective and industrially heterogeneous. It's dynamic, attracting new people, ideas, and energy by virtue of its own growing track record. The newest comers extend the story, and that's why students have always been an essential ingredient of this climate.[55]

In 2005, with a $35 million donation from German software entrepreneur Hasso Plattner, Stanford scaled up the model with the founding of the d.school. The "d" stood for "design" but also evoked the radical concept of "de-schooling." The organization's first annual report quoted an enthusiastic student who boasted that "I learned more about innovation in a week at the d.school, than I did in a year in my other classes."[56]

CMU: THE UNIVERSITY AS RENEWAL ENGINE

In contrast to MIT and Stanford, which promised rebirth from their privileged positions, CMU was at ground zero of American industrial decline, its fortunes tied to a city where unemployment in 1983 reached 18 percent. CMU's faculty and administrators experienced the competitive crisis as an acute reality and positioned themselves to offer solutions.

From the 1970s into the early 1990s, CMU developed its innovation culture out of four clusters of differing expertise. First, CMU was among the earliest universities in the United States to address industrial productivity as a scientific problem. In 1972, CMU engineers created the Processing Research Institute (PRI), an NSF-sponsored center to help mature industries confront the nation's slow productivity growth.[57] PRI brought engineering fields together and trained PhD students to focus on problems for Pittsburgh's corporate giants including Du Pont, Westinghouse, and National Steel. Neither "glamorous or exciting," CMU and NSF identified the program as a basis for future manufacturing research.[58]

A second foothold of technology–economics expertise was the Graduate School of Industrial Administration (GSIA). Founded in 1949, GSIA aimed to produce "an industrial 'renaissance man'" by bringing scientific rigor to business management.[59] GSIA's leading renaissance man was Herbert Simon, a polymath whose work spanned psychology, economics, organizational theory, artificial intelligence, and philosophy. But GSIA's experts also included AI-pioneer Allen Newell and computer scientist Richard Cyert, who would become the university's president and chief innovation evangelist.

A third locus of innovation expertise emerged out of CMU's industrial design heritage. In the 1940s, the university's predecessor, Carnegie Tech, offered the first industrial design degree in the United States with a faculty of craftsmen, painters, and illustrators training students for the Pittsburgh economy. In the late 1960s, this arts-oriented program remained on the margins of CMU's engineering science identity. Meanwhile, Simon and systems engineer Richard Felver developed a parallel vision of "design research" as the science of innovation.[60] In 1974, with half a million dollars in funding, CMU established the Design Research Center (DRC) uniting computer scientists, engineers, architects, and managerial theorists. DRC's founders promised a blockbuster payoff. "Our goal very simply put," they boasted, "is to increase the design capability of a practicing engineer by tenfold or more."[61]

CMU developed its fourth cluster of innovation expertise in its School of Urban and Public Affairs (SUPA), which integrated architecture, environmental policy, and urban planning. Founded in 1968, SUPA's approach to innovation again emerged from Pittsburgh's local conditions where waves of industrial boom and bust had created ethnic enclaves and racial tensions. SUPA's faculty (whom *Ebony* magazine called "SUPAstars") aimed to revitalize Pittsburgh through systems models, Black entrepreneurship, job training, and public–private collaborations.[62]

In the 1980s, CMU's clusters of innovation expertise converged through cooperation and competition. A key impetus was an NSF ERC award that infused $25 million into productivity-driven design.[63] Built on the foundation of the DRC, the renamed Engineering Design Research Center (EDRC) grew to over seventy-five faculty and graduate students, who developed futuristic manufacturing technologies, including wearable computers and 3D printers.[64] EDRC attracted sponsors from DEC to Exxon, Ford, and Boeing, and became an international site of pilgrimage for innovation policy experts looking for solutions to the competitive crisis.[65] Meantime GSIA's and SUPA's missions merged when CMU became a leader in two initiatives to restore Pittsburgh's manufacturing vitality. The first, the Ben Franklin Technology Center of Western Pennsylvania, sought to foster high-tech manufacturing start-ups that could generate jobs and economic growth. GSIA helped establish the second, the Enterprise Corporation of Pittsburgh, which also aimed to aid local entrepreneurs.[66] By 1993, the Enterprise Corporation claimed to have helped quadruple venture capital funds in the region, serve over one thousand clients, and spawn 2,500 jobs.[67] For its part, GSIA reported that 40 percent of its graduates were involved in start-ups that ranged from medical device manufacturers to natural-food restaurants.[68]

By the 1990s, a multidisciplinary ideal of innovation engaged nearly every unit at CMU.[69] University leaders pointed to research budgets, alumni data, and spinoff statistics to celebrate its unique innovation culture. The real proof, however, was out on the quad.

ROCKSTAR INTELLIGENTSIA

In the 1950s, Everett Rogers had discovered the characteristics of innovators by observing hardworking farmers in the heartland. Forty years later, CMU's Richard Florida found them playing Frisbee at an elite urban university. In his blockbuster hit, *The Rise of the Creative Class*, Florida recounted a campus stroll on which he "came upon a table filled with young people chatting and enjoying the spectacular weather." The group was there for more than just fun. Many of the Frisbee players, in matching T-shirts, were from an Austin-based software firm on the hunt for talent. The GSIA professor spotted "an obvious slacker" lounging on the grass with "spiked multicolored hair, full body tattoos and multiple piercings." To his surprise, Florida discovered that the slacker had just signed the highest-paying job offer in his department's history. "We want him," the young recruiters explained, "because he's a rockstar."[70]

Not since *Diffusion of Innovations* had one book so thoroughly integrated the spectrum of meanings about innovation and done so by putting innovators at the center. Published in 2002, *The Rise of the Creative Class* linked economics, geography, sociology, history, cultural studies, and personal biography in a wildly successful but deeply polarizing book that stands as the definitive statement of innovation culture. In contrast to the productivity indices of the technology–economics experts, Florida started with people. Ultimately, however, his work asserted that the rockstar and the productivity index were one and the same.

Florida was himself a child of the competitiveness crisis. The son of an Italian immigrant who had quit school at fourteen to work in a factory, Florida attended a Catholic school in his working-class New Jersey suburb. A gifted student, Florida received a scholarship to attend the agricultural branch of Rutgers University where he earned a degree in political science in 1979. Florida stayed on at Rutgers as a research associate on a federally funded study of the decline of American cities. He helped to collect and analyze six hundred interviews with Newark tenants and landlords. The data painted a bleak future for cities like Newark that had degraded infrastructure, widening inequality, a fleeing middle class, and an inability to compete in the global economy. These cities had maxed out their tax bases, and, after two decades of cash infusions, the federal government was withdrawing support. Indeed, Reagan's Department of Housing and Urban Development commissioned the study to find an excuse to do just that.[71] In 1983, however, Florida left Rutgers for Columbia, where he argued in his dissertation that deregulation was the real culprit responsible for urban decline. Florida traced how big banks had worked over a period of sixty years to privatize housing investment and he criticized the Reagan administration for continuing the trend.[72] While working on his dissertation, Florida taught at Ohio State University's Department of City and Regional Planning, where he rose rapidly from instructor to associate professor.

In 1986, the twenty-eight-year-old Florida came to CMU to expand his interdisciplinary approach to innovation policy. Working with his former Ohio State colleague Martin Kenney and their student Andrew Mair, Florida helped provide empirical data to heated debates about Japanese "just in time" manufacturing. The team mapped the location of Japanese auto plants on American soil and documented their impact on jobs.[73] In another project, funded by the Department of Commerce, Florida and Kenney charted the history and geographic clustering of venture capital. Debunking conventional wisdom, they demonstrated that there was no "specific recipe of factors" for venture success. They discouraged speculators from thinking that Boston and Silicon Valley could be reproduced and were cautious about public venture capital because of the risk of political interference and backroom deals.[74]

These geographic studies provided a foundation for a wholesale critique of the innovation policy consensus. In *The Breakthrough Illusion: Corporate America's Failure to Move from Innovation to Mass Production*, Florida and Kenney chided MIT's Commission on Industrial Productivity for its faith in technological breakthroughs. They argued that the technology–economics experts failed to offer a positive vision of "what the future should be" and pursued policies that hurt industrial workers and the poor. The pair placed their hope in repairing the decades-long separation of innovation from production with a focus on labor rather than technology.

Florida claimed that work in a knowledge-based economy was undergoing historic change that generated a fundamental dilemma. A distinguishing feature of the new economy was the "hypermobility" of "high-tech think-workers," which conferred great benefits on those with the right talents. Yet, hypermobility came with major social costs. Egged on by venture capitalists, companies raided each other for talent. Fewer products were commercialized due to the instability. Entire regions lost out from brain drain. The biggest problem, however, was a widening gap between knowledge work and manufacturing. In high-tech industries such as semiconductors, most production workers were women and minorities who were treated as "second-class citizens," subject to low pay, poor working conditions, job insecurity, non-union jobs, offshoring, and subcontracting. To "tap the energy" of people at risk of being left behind, the pair argued for the conversion of military spending to civilian commercialization, the fusion of R&D with manufacturing, and training to help all workers to become think-workers.[75]

Pittsburgh was Florida's testbed for reforming the knowledge economy. In 1993, Florida became the director of the Center for Economic Development (CED), yet another cross-disciplinary unit at CMU. One of CED's projects was the establishment of Pittsburgh's Regional Economic Revitalization Initiative, which promised to give all Pittsburgh's stakeholders an opportunity to work together toward a better alternative to the region's status quo. Engaging 5,000 stakeholders, the initiative claimed that its participatory process was at least as important as any specific results it might produce. With technocrats and special interests running the initiative, however, the revitalization project reproduced many of the shibboleths of innovation policy. It employed Saxenian's principle of regional advantage. It also advocated for continued investment in robotics, biomedical research, and environmental technology, arguing that funding advanced research would spawn well-paying manufacturing jobs. Achieving success, however, required "cutting-edge" labor-management partnerships, tax breaks, and modernized government services with "a customer focus." Finally, to win the competition for hypermobile workers, Pittsburgh needed better nightlife and dining (which lagged behind revitalized "entertainment zones" in Cleveland and San Antonio).[76]

As Florida implemented his ideas locally, his national profile grew. In 1996, his ideas reached the highest levels of innovation policy when President Clinton asked Florida to work with Branscomb and others to find a bipartisan innovation policy consensus.[77] The convening assessed the efficacy of SBIR and compared different models for university technology transfer. However, Florida, like Rogers before him, was after a bigger synthesis, one that explained why the tattooed rockstar on CMU's quad was "another talented young person leaving Pittsburgh."[78]

INNOVATORS OF THE WORLD, UNITE!

Innovators were once thought to be a rare breed, no more than 2.5 percent of the population. However, as high-tech entrepreneurship proliferated, government investment flowed, and university programs flourished, the relative proportion of innovators to the rest of the population was growing. "Capitalism," Florida opined, has taken "people who would once have been viewed as bizarre mavericks operating at the bohemian fringe and setting them at the very heart of the process of innovation and economic growth." Innovators had become a bona fide economic and social class.

Florida's demographic case for a new class of think-workers was strong.[79] Since the late 1970s, charts of the shifting fortunes of the manufacturing and information sectors had circulated among policy experts.[80] Florida broadened the analysis to include labor writ large. Agricultural workers had vanished in the twentieth century, from over one-third of all American workers to less than 1 percent. The industrial working class had peaked around World War II at 40 percent and was in steady decline. Two categories of work, however, had grown dramatically. Service jobs tripled, making up over 40 percent of the nation's workers, and, "creative class" jobs grew at the same magnitude but with a dramatic uptake after 1980. Florida defined the creative class as any group of workers who regularly used creativity to solve problems in new ways. It included a "super-creative core" of scientists, engineers, artists, writers, entertainers, university professors, and thought leaders. These high-skilled workers had ascended fastest, jumping from 2.4 percent of the population in 1900 to 12 percent in 1999.[81]

Economic data measured by the now robust national innovation system reinforced Florida's conclusions. Despite the anxieties of the 1970s, investment in R&D had steadily increased since World War II. The regulatory reforms of Presidents Carter and Reagan meanwhile had resulted in a doubling of patents filed in the 1990s.[82] According to Florida, these gains stemmed from "the creative furnace inside each human being." That meant that creative workers were themselves "the means of production."[83] Echoing Stanford's design experts, Florida contended that *everybody* had the inherent capacity to be creative under the right conditions.

Expanding on the work of the innovation geographers, Florida argued that there were critical ingredients to a region's innovative capacity, which he called "The 3T's." The first, *technology*, was the defining enterprise of the

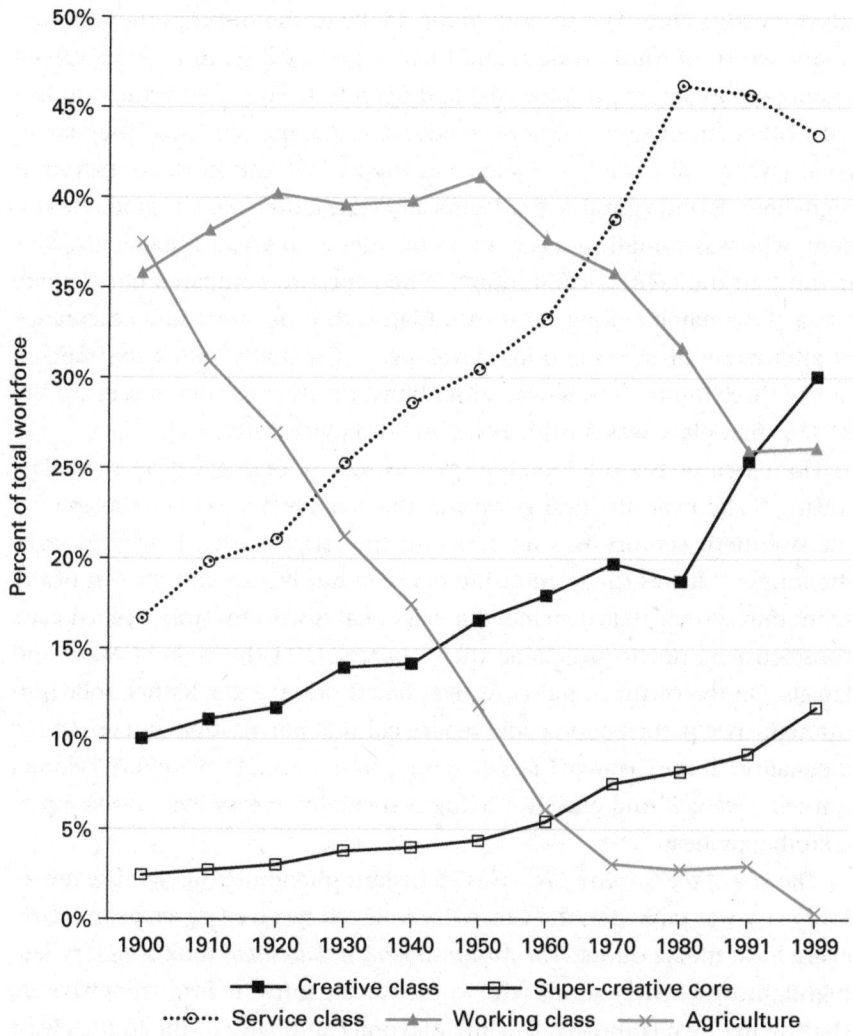

FIGURE 6.3
The class structure as redefined by Richard Florida; chart recreated with US Bureau of the Census data reported in *The Rise of the Creative Class*.

super-creative core. The second, *talent*, spoke to the disproportionate economic worth of think-workers and their hypermobility. Innovators valued cosmopolitanism, merit, diversity, and openness. They also wanted to live with other innovators. No longer bound to company towns, they could go anywhere. The third, *tolerance*, was the newest and most controversial ingredient. Florida got the idea from Gary Gates, then a CMU graduate student, who was compiling a Gay Index of cities with socially liberal lifestyles friendly to the LGBTQ community. When the two compared notes, there was a close match among cities with high-tech innovation and acceptance of alternative lifestyles. Florida developed a Creativity Index that ranked nearly three hundred American cities based on their relative success in the 3T's. In first place was Austin, Texas; in 90th place, Pittsburgh.

Florida emphasized a looming crisis as society underwent a "great class sorting." The near identical growth of the creative and service classes over the twentieth century was no accident; the latter existed largely to serve the former.[84] It was the relationship between innovators and the rest of the economic system that demanded a class analysis. Innovators needed class consciousness not to overcome their oppressors in the style of Marx and Engels. On the contrary, most creatives had it pretty good. Rather, solidarity through civic participation could reduce cultural polarization and economic inequality. "If the Creative Class does not commit itself to this effort," Florida warned, "we will find ourselves living perpetually uneasy lives at the top of an unhappy heap."[85]

The Rise of the Creative Class was an instant phenomenon. Because nearly every city was ranked in the Creativity Index, it received coverage in nearly every local media outlet. *The Austin American Statesman* took a victory lap, highlighting a CMU alumna who worked at the software firm Trilogy before starting her own company. Dueling mayoral candidates in the Index's least creative city, Memphis, Tennessee, claimed that they needed "to be more of a union than a confederacy" and that they had "the recipe for success," they just needed "to cook it." City planners in Minneapolis worried that their once innovative anchors like 3M were now beholden to Wall Street and dragging the city's reputation down. "Hold on to your newspaper" the *Kansas City Star* warned social conservatives as it explained the link between high-tech workers and the Gay Index. West Virginia's *Charleston Gazette* used Florida's study to call for environmental protection; "how many software engineers," it editorialized, "want to move to a place known for shearing off its mountaintops?"[86]

The Rise of the Creative Class was also a lightning rod for criticism. Geographers questioned the link between bohemianism, tolerance, and growth, arguing that what creative people really wanted was lower taxes, less regulation, and better schools. Early critics of neoliberalism called the creative class a fantasy designed to "publicly subsidize . . . a circulating class of gentrifiers, whose lack of commitment to place and whose weak community ties are perversely celebrated." Scholars at the conservative Manhattan Institute argued that the creative class thesis was an exercise in circular logic—vibrant neighborhoods were the result, not the cause of economic growth. Social conservatives attacked the link between innovation and tolerance as a sign of elitist decadence. One *Sacramento Bee* subscriber railed against the paper's praise of Florida's ideas as "a creative case of prolonged psychobabble . . . a good example of the liberal agenda teaching in most universities today." Another warned, "How about not accepting Sodom and Gomorrah? And wasn't it the creative class that caused the worldwide flood?"[87]

Whether enthusiastic or dismissive, most commentators glossed over Florida's message about a class war. They could be forgiven the oversight. Despite the service class's centrality to the economy, its plight received disproportionately small treatment. Moreover, Florida offered little advice on how all jobs could become creative jobs.

One undeniable outcome of *The Rise of the Creative Class* is that it made the innovation expert into a rockstar. The book helped Florida become one of the world's most sought-after consultants. In 2003, he headed the Memphis Manifesto Summit, the first official creative class gathering with a "Creative 100," nominated by their peers for innovative talent.[88] The following year, Florida moved to the Brookings Institution in Washington, DC. He later made a home in Toronto where he formed the Creative Class Group, a consultancy whose clients included cities throughout the world and companies and organizations that range from Meta to Cirque du Soleil.

Florida had become another talented person to leave Pittsburgh.

CHANGE OR DIE!

By the late 1990s, the demand for innovation was a national imperative sustained by a federal network that linked programs in nearly every state and city in the United States. It was the rare politician who did not enthusiastically promote a pro-innovation platform. Public funding and private capital ballooned to tens of billions of dollars aimed at generating technology, businesses, jobs, and healthy communities. University initiatives to cultivate innovators and entrepreneurs expanded from a few hundred students and faculty in the early 1970s to tens of thousands annually.

A diversified cadre of experts representing different generations, subspecialties, and political commitments helped bring this imperative to life. Technology–economics experts who had come of age during the Cold War developed an industrial policy that portrayed innovation as a problem of and a solution to productivity. Regional innovation experts used geography and anthropology to argue for the power of local culture. Advocates for entrepreneurship expanded incubators to foster high-tech ventures and create manufacturing jobs. Design experts emphasized human-centered engineering to build an empowering digital future. Urban planners applied innovation-based methods to foreground the processes of community-driven change. This proliferation spawned different, sometimes competing, interpretations of what counted as innovation and what strategies were necessary to achieve it. Florida was the imperative's most influential alchemist. His interpretation of a world transformed by technology, talent, and tolerance did as much as any federal policy to scale up the innovation imperative by unifying the seemingly opposing poles of technology–economics and society–culture.

This merger of productivity and creativity comes into focus when viewed across the long arc of the competitiveness crisis. As the MIT Commission on Industrial Productivity demonstrated, innovation policy ascended because its deficits could be measured from the factory floor to the global productivity index. The creative class performed a similar trick of scale, unifying individual, regional, and national identity. Together, the technocrats and the champions of bohemia gave a name to phenomena that once described could be found everywhere and was nearly impossible to unsee.

At the same time, both industrial policy and creative class consciousness brought uncomfortable realities of the innovation imperative into view. Were precarity and inequality an inherent consequence of a knowledge economy? What would it mean for service jobs to become innovation jobs? How could tolerance and diversity be made more than lip service in a culture that was predominantly male and white? The stakes were high. If all Americans were not incorporated into innovation culture, the United States could fall behind. Or worse, crack apart.

7 EMPOWER EVERYTHING

For change to be a way of life rather than an occasional traumatizing shock, the "Indians" as well as the "chiefs" have to be engaged in change making and change mastery—while still doing their necessary jobs.
—Rosabeth Moss Kanter, 1983

Our mission is to empower every person and every organization on the planet to achieve more.
—Microsoft, 2022

Roberta Briggs was in the middle of an innovation breakthrough. In 1977, when she initiated the change, her company Chipco was at a crossroads. The microelectronics producer had made celebrity millionaires out of its founders, but fast growth was in the past. Briggs was a lowly personnel manager at a manufacturing plant in Boston's outer suburbs. The facility had over one thousand employees, mostly women and immigrants engaged in batch assembly. Her management peers were all men who corporate headquarters deemed unpromotable. Chipco once had a reputation as an egalitarian company in which anyone with talent could advance. Then, on a site tour, a company MBA let slip that the workers looked like "caged animals."

Facing productivity losses, Chipco was about to implement assembly-line techniques. Fear of layoffs was palpable. Briggs was sure that the standardization would lead to unrest and decline. She also knew that Chipco's workers had the intimate knowledge to rescue the company, but they needed to be treated as valued partners. They needed to find meaning in their labor. They needed to be *empowered*.

Briggs had a blueprint for change. She was inspired by a hot-off-the-press feminist critique of corporate life, *Men and Women of the Corporation* by sociologist Rosabeth Moss Kanter, that explained the plight of minority "tokens" inside organizations.[1] In an experimental program, Briggs brought together her reluctant supervisor with her skeptical workers to redesign a new production process. Her intervention avoided the assembly line and enhanced the bottom line.

Briggs had one more tool at her disposal: Kanter herself. Chipco, a pseudonym for a major American technology corporation, had hired the sociologist's consulting firm, Goodmeasure, Inc. to reverse its sputtering productivity. According to Kanter, "entrepreneurial middle managers" like Briggs were transforming America's institutions from within. The contributions of these quiet innovators revealed a powerful new truth: to be more innovative, companies needed to become more equitable.[2]

This chapter investigates the rise of expertise in inclusive innovation. Since the late 1960s, experts had worked to increase the nation's innovative capacity with an image of jet-setting innovation millionaires. This ideal did not apply to an employee like Briggs, removed in time, sex, and authority from founders' myths. Nor did it address the square middle of employees who were not obviously members of the creative class. Moreover, as economic geographer Richard Florida's research would later reveal, there was a negative correlation between the concentration of high-tech firms and nonwhite populations.[3] In image and reality, large segments of Americans had been excluded from the innovation economy.

In the 1980s, experts sought to broaden participation in innovation along two distinct paths. First, consultants promised large corporations that they could become more competitive by capitalizing on the promise that *any* employee could become an internal innovator. Second, reformers worked to enhance access for women, Black Americans, and other minorities to careers in science, engineering, and business.

By the 1990s, innovation experts, politicians, and diversity managers fused these approaches into the now sacrosanct belief that *diversity drives innovation*. This merger between feminist and antiracist principles and the innovator's toolkit was neither inevitable nor accidental. It was first forged by Kanter, a sociologist with seemingly disparate lives, first as a pioneering feminist scholar, and then as a top innovation guru. A wide range of corporate diversity initiatives, professional associations, and state and federal policy programs build upon this vision of empowerment. These advocates navigated transformations in the American workforce and the shifting politics of gender and race with the goal of changing the face of innovation culture.

BARRIERS TO INNOVATION

Since the 1960s, experts have argued that innovation is both transgressive and inclusive.[4] Early innovation experts such as Donald Schön and Warren Bennis drew inspiration from social justice movements in their models for remaking American institutions. They cast innovation as a progressive force that bridged political interests and integrated science, technology, and society. Their theories emphasized viewpoint diversity in fostering innovation and supported the conclusion that disruptive change came from thinking differently.[5]

In practice, innovation culture was defined by white men. The first generation of innovation experts were all men. The technoscientific domains in which they applied their ideas still largely barred women and minorities from entry. The heroes of magazines like *Innovation* were hard-working and hard-partying male competitors; when women appeared in their pages, it was as secretaries, wives, mistresses, or muses. The first university entrepreneurship programs that these experts helped build disproportionately attracted male students. Some immigrants found success in these initiatives, but Black and Hispanic Americans were almost entirely absent.

In the 1970s, women made hard-earned gains in American society in general and in technoscientific careers in particular. Legislation including Title IX and the Women's Educational Opportunity Act opened access to elite colleges and universities. Large corporations, which had segregated women's work into feminized staff positions, began to hire women into managerial roles.[6] Still, the gap to equal participation in science and technology was large, and progress was slow. In 1978, women accounted for just 20 percent of scientists and 2 percent of engineers and the disparity was especially stark in higher levels of management and administration.[7]

Recruitment initiatives defined women's participation in technoscience as a creative force for social good. A 1974 campaign for high school students by the Engineers' Council for Professional Development (ECPD), for example, asserted that women engineers were part of an "awakening movement." ECPD profiled women studying and working, juxtaposed with illustrations of circuits and butterflies, and promised that women in technical fields would help "reclaim wasted land, clean the air, and make the world a better place to live in." Women would encounter stereotypes that engineering was "intrinsically masculine," but their success would help make entry into the profession easier for the next generation.[8]

Feminist critics and reformers meanwhile addressed ongoing barriers to participation in the innovation economy. Workshops, conferences, and university courses explored the roots of inequality and outlined strategies to combat discrimination and internalized attitudes of inferiority.[9] Historians of science and technology investigated women's past challenges and achievements, while sociologists advocated for improvements such as flexible work time and family leave.[10] More radical critics, including Sally Hacker, described engineering culture as a misogynist priesthood in need of radical change. "The argument is not that the mere presence of women will somehow magically humanize these institutions," she wrote, rather "it is in these highly significant institutions that direction for change takes place, and where we can also see the sharpest differential allocation of power between men and women."[11]

Black Americans made similar gains in the innovation economy in the Civil Rights era. Affirmative action programs opened scientific and engineering programs to Black students from which they had been systematically excluded, while new advocacy groups including the National Society for Black Engineers (NSBE) and the National Advocacy Council for Minority Engineers (NACME) provided mutual professional aid.[12] Black scientists and engineers began to find employment in federal agencies and large corporations including Fairchild Semiconductor, Hewlett Packard, and IBM, which had comparatively robust equal opportunity programs. Progress, however, was even slower for Black Americans than for women. After more than a decade of affirmative action and equal opportunity programs, just 1 percent of the nation's scientists and engineers were Black. Those employed in computer specialties amounted to a minuscule 0.2 percent of that 1 percent.[13]

Efforts to expand Black participation in technology stressed entrepreneurship and economic self-determination to a greater extent than

programs aimed at women.[14] Federal interventions targeted business formation, including 1965 Executive Order 11246, which required nondiscrimi-
nation in federal contracting, and the Department of Commerce's creation
of the Office of Minority Business Enterprise.[15] Additionally, a few high-tech
firms supported "Black capitalism" by providing business, equipment, and
mentoring to minority-owned subcontractors such as the Hewlett Packard–
supported East Palo Alto Electronics and the GE-backed Progress Aerospace
Enterprises.[16]

A key strategy for enhancing participation was an emphasis on correcting the "hiddenness" of Black innovators. Reformers used history as a tool
for creating counter-myths to narratives of inferiority by demonstrating
that Black Americans had made major contributions as scientists, engineers, inventors, and entrepreneurs in the face of exceptional obstacles.[17]
A series of books in the 1970s uncovered "hidden contributors" and "Black
innovators" from Benjamin Banneker at the founding of the republic to
Elijah McCoy, Lewis Latimer, and Garrett Morgan in the age of American
invention.[18] Additionally, new media outlets such as *Black Enterprise* showcased contemporary successes.

In parallel to this integrationist approach, Black liberation activists portrayed science and technology as tools of oppression but also as possible
sources of power. Washington University's Robert C. Johnson, for example, challenged the dominant frame of the "Wonders of the White Man's
Science and Technology" and the disproportionately negative impacts of
technological change on Black communities. He and other critics sought a
scientific and technological identity that could "permeate our personal and
social development as Black people."[19]

Despite occasional uses of the term innovation, efforts to increase participation of women and minorities in technoscientific careers in the 1970s
did not emphasize the concept as a guiding framework. With a few note-
worthy exceptions, the immediate goals of reformers were to create opportunities for women and minorities in science, technology, and business and
to help them fit in. Conversely, among innovation experts there was almost
no attention to the challenges faced by women and minority innovators.

To understand how gender and race became matters of concern in innovation culture, we need to examine the career and ideas of the expert who
made the largest contribution to their coevolution.

SEEKING UTOPIA, FINDING TOKENISM

Rosabeth Moss was born in 1943 in the middle-class Jewish community of Cleveland Heights, Ohio. Her mother was a teacher and her father a lawyer. She was a talented writer who penned a mystery novel at age eleven and won a statewide journalism contest in high school.[20] At a time when some elite universities still barred women, Moss attended Bryn Mawr where she graduated with high honors. During her junior year, she married Stuart A. Kanter, a student of social psychology. The two attended graduate school at the University of Michigan, where, driven by an interest in alternative futures, Kanter focused on nineteenth-century communes. She completed her PhD in sociology in just three years, and joined the faculty at Brandeis University while Stuart began teaching at the Harvard Business School.

In March 1969, tragedy upended Kanter's life when her husband died by suicide.[21] That summer, Kanter joined the Cumbres human potential commune in New Hampshire as both a member and a participant observer. Cumbres, led by former Cuban revolutionary Cesareo Pelaez, was one of dozens of countercultural experiments taking place across the United States. The following year, Kanter took up residence in a different commune in Bethel, Maine.

In search of "new ways of being and doing," Kanter combined her dissertation work with her lived experience. She wanted to know how goals of social equality were or were not met by the communal structures used to achieve them. How, in other words, did communes attempt to innovate? Her research questions hinted at her later appeal as a consultant: "What makes some communes work, while others fail? Why is it so difficult to put utopian ideals into practice?" Kanter found the answers in everyday tasks. She documented how communes confronted practical problems such as how to assimilate new members, how to preserve individuality in a collective, and how to become self-motivated to do necessary labor. "In Utopia," she quipped, "who takes out the garbage?"[22]

Kanter also investigated a range of approaches to sexual relations. The Shakers divided men and women into separate spheres in pursuit of celibacy. Virginia's Twin Oaks community in contrast made no distinction between sex roles when assigning jobs and members referred to each other with the gender-neutral pronoun *co*.[23]

When the countercultural movement collapsed, Kanter transitioned from studying communes to corporations. She began with a critique of her own field, arguing that organizational sociology's masculine ethic stemmed from corporate demands for order. This authoritarian foundation was epitomized by the "scientific management" of efficiency expert Frederick Taylor.[24] According to Kanter, subsequent reformers such as Elton Mayo had introduced collaboration and worker participation into human relations but maintained applied sociology's paternalist authority. Kanter asserted that feminist approaches could aid in remaking power relations in large bureaucracies by analyzing how corporations structure the lives of those who worked within them. She was especially concerned with how women and minorities fared as they sought to enter the managerial ranks. The answers were not good.

In 1977, after hundreds of hours of observation and interviews, Kanter published her monumental book *Men and Women of the Corporation*. Her analysis included differences of race and class inside corporations, but her primary focus was the experience of women. Her case study was the pseudonymous Industrial Supply Corporation (Indsco), a giant company with over 50,000 employees and a heritage dating to the nineteenth century. Like most corporations in 1970s America, Indsco was confronting the challenges of equitable work. Indsco described itself as a "people-conscious organization," yet women accounted for just 10 percent of the lowest levels of management and none of the higher ranks.

Men and Women of the Corporation popularized the term *tokens* to characterize the status of minorities working inside corporations. "The life of women in the corporation," Kanter concluded, "was influenced by the proportions in which they found themselves."[25] Kanter identified how structure and demography created negative social, personal, and career effects on tokens. Some, such as unequal pay, were easy to quantify. But the harms were also psychological and cultural. Kanter described how token women faced gendered stereotypes that included the helpful *Mother*, a service-class hero to whom male employees expressed their private troubles; the *Seductress*, who was a sex object; the *Iron Maiden*, who fought back against discrimination and often got "left to flounder on their own"; and the *Pet*, whom men viewed as a "cute, amusing little thing." Because resisting these stereotypes was a constant battle, women perversely came to enact them. These less quantifiable harms resulted in hostile work environments, high turnover, and failure.

Power was at the center of Kanter's analysis. For Kanter, power was not inherently a weapon of subjugation. It was "the ability to get things done, to mobilize resources, to get and use whatever it is that a person needs for the goals he or she is attempting to meet."[26] In the right system, power could be a progressive tool.

Kanter's purpose was not merely academic; she wanted to provide solutions that aided the lives of corporate women and minorities. She was critical, however, of self-help schemes that failed to attack the "root causes" of discrimination and thus offered "a subtle and insidious system-maintaining message."[27] She drew from her commune work to suggest interventions that could increase minority power inside organizations. First, to enhance opportunities, companies could implement job redesign, flexible working hours, and manager reeducation. Second, "empowering strategies" could give minorities a greater voice by "flattening the hierarchy," employing autonomous working groups, and implementing mentorship programs. Finally, the use of number balancing through batch hiring and cross-organization networks could combat tokenism.

GOODNESS EXPERTS

How did an academic critic of corporate capitalism become a top innovation expert? For one, *Men and Women of the Corporation* was as much a sensation among business executives as it was academic sociologists. In an era that saw the establishment of diversity officers in human resources departments and business ethics as an academic specialty, corporations looked to experts like Kanter for answers. Indeed, with her book's publication, Kanter accepted a prominent position at Yale's new School of Organization and Management.

Kanter's personal life also shaped her professional transformation. In 1968, at the start of her career, a friend had introduced Kanter to Barry Stein, an MIT-trained engineer and former Arthur D. Little consultant. Like Kanter, Stein believed in the practical potential of radical theory. In the early 1970s, he quit corporate consulting to apply his skills at a nonprofit center dedicated to the self-determination of Black and Native American communities.[28] The two married in 1972, and Stein returned to MIT to complete a dissertation that advocated for small-scale industries that fused communalism with the innovation expertise of Bennis and Schön.[29]

After the runaway success of *Men and Women of the Corporation*, Kanter and Stein established Goodmeasure, Inc., a consulting firm that promised to help corporations develop progressive policies and achieve diversity goals. Goodmeasure's value proposition was that companies with progressive human-resource practices were not just better places to work, they were more profitable. The firm's services included bias training for executive managers, programs to enhance the number of women and minorities in senior management, and work–life initiatives such as flextime. Kanter also established a "progressive index" that surveyed Fortune 500 companies to create a buzz about which companies had the best human relations policies.

Goodmeasure's first breakthrough product was *A Tale of "O,"* an animated film that dramatized the discriminatory perils of tokenism.[30] Goodmeasure used *A Tale of "O"* in training seminars and sold the video to Fortune 500 companies, universities, professional societies, NGOs, churches, hospitals, and government agencies in the United States and abroad, accompanied by guides for role-playing and strategic planning.[31] The film built on *Men and Women of the Corporation*, in which Kanter had illustrated how difference stood out via a simple abstraction:

$$X \, X \, x \, x \, X \, X \, O \, X \, x \, X$$

In a series of workplace vignettes, *A Tale of "O"* gave life to X's and O's to show the discrimination that tokens encountered in their organizations, explaining the problems of "visibility," "contrast," and "assimilation."

VOICE: Could you speak at our next program? We're devoting it entirely to the problems of O's.

Eventually, if the O does well it will be expected to meet with every board or task force or planning group.

VOICE: I really think an O ought to be included.

VOICE: But we don't have any other qualified O's.

For the O, this often results in overload.

VOICE: I guess those O's just can't take the heat.[32]

According to Goodmeasure, the use of abstract symbols made diversity training more palatable in conservative boardrooms and avoided placing unwanted attention on minoritized individuals.

As Goodmeasure's portfolio of clients grew, the firm developed a theory of "parallel organizations" to address the structural failures that Kanter identified in *Men and Women of the Corporation*.[33] Parallel organizations were "flat, flexible, but formal problem-solving and governance" units that could be built inside large bureaucracies. The promise was that parallel organizations could enhance problem solving in environments of high uncertainty while also helping workers to access "opportunity and power" that the bureaucracy denied them.[34]

Goodmeasure's proof of concept came at Digital Equipment Corporation (DEC), which Kanter later pseudonymized as "Chipco." Since its founding by entrepreneur Ken Olsen in 1957, DEC had served as the source of many of innovation culture's myths. Kanter's informants described their company as "a gypsy society, a university, a theocracy, twenty-five different companies, and a company with 'ten thousand entrepreneurs.'"[35] However, in the 1970s, DEC was struggling with bigness and global competitiveness.

In 1977, a personnel manager named Fran Grigsby (the pseudonymized "Roberta Briggs") convinced her company to hire Goodmeasure. The firm diagnosed DEC's inability to adapt as a problem of limited employee access to power and introduced a parallel organization in which assembly workers and senior managers seven levels up the chain of command collaborated. The committee in turn established a charter in which anyone at DEC's assembly plant had the right to "participate in the experimentation of managing change." Pilot teams then spent six months reimagining the production process. After implementing changes, the teams dissolved. Goodmeasure concluded that participants developed new skills, were more productive, and had higher job satisfaction. Further proof of success was the widespread raises and promotions across company ranks.[36] Left unspoken was that the intervention avoided talk of a worker's union.

Kanter's DEC consulting was her first foray into the innovation business. After the project, Goodmeasure took a deep dive into innovation expertise, identifying ninety-nine theses about how innovation worked.[37] The top finding: the organization made the innovator.

QUIET INNOVATION

It was one thing for an entrepreneurial outlier like DEC to foster commitment and community among its corporate workers. Could the lumbering firms and failing industries that critics blamed for American decline also adapt to the age of innovation?

In 1983, Kanter incorporated lessons from consulting at more than fifty companies into a manifesto for how reformed bureaucracies could enhance national competitiveness. Her book, *The Change Masters: Innovation for Productivity in the American Corporation*, examined "traditional" organizations struggling to change and "new style" companies at the forefront of innovation.[38] Kanter began by explaining the many ways in which companies tried, but failed, to innovate. She then described successful innovative environments, focusing on their use of parallel organizations. She followed with a set of reforms to aid managers in innovating inside their own companies.

The openly feminist mission of Kanter's early work was subdued almost to the point of absence in *The Change Masters*. Many of the employees she profiled, however, were minoritized women middle managers in "ghettoized job assignments" in areas of purchasing, personnel, and public relations. *The Change Masters* also emphasized diversity in perspective, job roles, sex, race, class, and ethnic background as a part of what made teams more innovative than individuals. Finally, Kanter concluded that a true "corporate Renaissance" demanded replacing "managerial superiority, male superiority, white superiority" with egalitarian participation.[39]

The Change Masters's central argument followed directly from *Men and Women of the Corporation*—the most innovative companies were those that empowered their employees. Under the right conditions, every individual inside that system had the potential for innovation. In a hierarchical organization, however, subordinates could not rule by force. They needed to be *granted* power. Companies that were "innovation stimulating" were those that deliberately distributed power, while those that withheld access to power were "innovation smothering."

The Change Masters dispelled heroic myths of lone innovators and valorized the hidden work of middle managers like Briggs. Kanter was not the first to note the importance of these "intrapreneurs" nor was she their only champion.[40] But Kanter gave intrapreneurs the first systematic treatment.[41] Kanter argued that intrapreneurs shared common skills and pathways with their flashier entrepreneur cousins. Like all innovators, intrapreneurs were visionaries. Because they were positioned in the middle, however, their ability to achieve their goals was dependent upon navigating bureaucracy. They could not be "rugged individualists," rather, they had an "integrative mode of operating" as team builders and cheerleaders.[42] Their temperament was balanced and empathetic. Briggs, for example, was "known for walking through the plant almost every day" and spent "her time actively listening to the cues about coming problems."[43]

Corporate change masters rarely produced breakthrough discoveries or system-altering changes. They were "authors not of the grand gesture but of the quiet innovation." Their contributions consisted of "applying ideas that have proved themselves elsewhere," or "rearranging parts to create a better result," or "noting a potential problem before it turns into a catastrophe."[44] Collectively, however, the incremental innovations of change masters amounted to system-changing improvements.

Kanter offered a series of *power tools* to establish the environments in which corporate innovation could flourish. Power tools included access to knowledge and training as well as resources that gave employees time, space, and permission to pursue new ideas. Kanter argued that power tools should be accessible to anyone at any level in an organization. To provide that opportunity, companies needed to create teams that cut across hierarchies. Once silos were dismantled, an organization would benefit from "the energy and excitement that are unleashed, the ideas that come bubbling forth, the zest for work that suddenly appears in employees who had looked like 'deadwood.'"[45]

THE CHANGE MASTER

Kanter's prescription for a corporate Renaissance launched her to the top rank of innovation gurus. By the mid-1980s, Goodmeasure had annual revenue of $5 million and served hundreds of clients, including Apple, DEC, GE, GM, Honeywell, IBM, and Xerox. Meanwhile Kanter charged $15,000 per speaking engagement. In 1986, she was recruited to the Harvard Business School, which prior to her arrival had only one tenured woman faculty member and was embroiled in a sex discrimination lawsuit.[46] Three years later, she took over as editor-in-chief of the *Harvard Business Review*.

Kanter's success put her in the room during some of the tech industry's biggest shakeups, which she characterized as growing pains of a "post-entrepreneurial" world. Her 1989 book, *When Giants Learn to Dance*, explained that corporations were finding their way between the valorization of digital innovators and stable productivity. To prove the point, Kanter described her consulting work for Apple Computer. When she arrived at Apple on a stormy November day in 1984, she found a company in turmoil. A freewheeling culture led by a mercurial founder had created an untenable environment that was "not liberating but curiously disempowering." From her insider's vantage, Kanter chronicled Apple's "apocalyptic change," including Jobs's infamous ouster. In the wake of Jobs's firing, Kanter championed Apple's growth from its glitzy entrepreneurial days to a new era in which the "heroes" were teams rather than hackers and engineers.[47]

As Kanter's celebrity grew, she was profiled in terms that challenged and reinforced stereotypes of working women. The *New York Times* described her as "A superwoman of sorts . . . a wife, a working mother, a Yale sociology professor with classes to teach."[48] The *Washington Post* declared Kanter "happiest when her appointment book is fullest," explaining that "she has a fax machine at her summer home on Martha's Vineyard and a car phone, and flies the Concorde to make quick turnarounds on European business trips." She nonetheless was a good mother who, "when school schedules permit[ted]," would take "her 10-year-old son, Matthew, along with her on business trips."[49] At the same time, her fame as a corporate champion sparked backlash among the feminist academics for whom she had been a role model. In a *Working Woman* profile, Kanter's own dissertation advisor concluded that "she seems to have lost her critical edge."[50]

The longer that Kanter spent in corporate consulting, the more her attention shifted from the middle to the top. She also had less time for the rigorous groundwork that had distinguished her from lesser prognosticators. In 1997, Kanter coedited the book *Innovation: Breakthrough Ideas at 3M, Dupont, GE, Pfizer, and Rubbermaid* in which technology executives promoted their companies with little mention of equity, diversity, or middle management.[51] Kanter followed in 2001 with *Evolve! Succeeding in the Digital Culture of Tomorrow*, a guide for managers that developed out of Martha's Vineyard conversations with the likes of Bill and Hillary Clinton, Alan Dershowitz, and Wolfgang Puck. The book began with "Evolve!—the song," which Kanter wrote herself after sharing the stage with Justin Timberlake at a White House philanthropy event. Her verse proclaimed that the internet was no place for quiet intrapreneurs:

Think big! Think bold! Think of new things to create.

First-rate prizes go to leaders who can innovate.[52]

FIGURE 7.1
Rosabeth Moss Kanter posing in front of Goodmeasure, Inc. with her bestselling
Change Masters for *Ms.*'s Woman-of-the-Year award. Copyright John Goodman 1985.

DIFFERENCE ENGINE

The twin demands of equity and productivity that Kanter fused into common purpose were at the forefront of American working life in the 1980s and 1990s even as her focus shifted to the C-suite. An array of consultants, diversity officers, academics, pundits, and politicians sought to enhance the participation of women and minorities in the innovation economy. Their efforts added the value of *difference* to innovation.

These inclusive innovation proponents operated under altered demographic, political, and cultural conditions than those that had inspired *Men and Women of the Corporation*. By 1993, women made up 45 percent of American workers, a doubling since 1960 and a quadrupling among younger women.[53] Female entrepreneurship had grown at twice the national average, while Black-owned businesses more than tripled from 1972 to 1992 with the greatest expansion in white-collar industries. However, women and minorities were still significantly underrepresented in technoscientific professions, particularly in managerial ranks.[54]

The largest demographic change in the innovation economy was high-tech immigration from Asia. By the early 1990s, foreign-born students earned 40 percent of the nation's science and engineering PhD degrees. Silicon Valley was the epicenter of high-tech migration with foreign-born professionals accounting for a third of the region's innovation workforce. The first cohorts of migrants experienced many impacts of tokenism including isolation, racism, and discrimination, while later generations confronted "model minority" stereotypes.[55]

Expert studies of gender and race in the workplace had also grown significantly since the 1970s. Kanter's insights remained the starting point for a new field of gender, work, and organizations.[56] As harassment and sexism persisted, however, scholars challenged the quantitative assumptions of tokenism and liberal feminism's assimilationist approach to power.[57] At the same time that Kanter was writing *The Change Masters*, for instance, a team of Marxist scholars argued that the microelectronics industry had a patriarchal division of labor that could only be broken down if "feminism is on every agenda."[58] In 1994, the National Research Council released a major report asking why there were still so few women in technoscience. It highlighted successful initiatives by which technical women had directly confronted paternalism, harassment, and their impact on attrition.[59]

Meanwhile diversity consultants and managers argued that support-
ing difference was both a moral imperative and a source of productivity.[60]
R. Roosevelt Thomas, Jr., founder of the American Institute for Managing
Diversity, insisted in an influential *Harvard Business Review* article that for
Black professionals to rise above entry-level jobs, corporations needed to
shift their philosophy from affirmative action to "affirming diversity."[61]
Santiago Rodriguez, Apple Computer's first director of multicultural programs, similarly asserted that "meaningful sameness and meaningful
difference" among women, Black, Hispanic, international, and LGBTQ
employees fostered "vitality, creativity, new ideas, and growth."[62]

Magazines such as *Black Enterprise, US Black Engineer,* and *Hispanic Engineer* were important sites of exchange among minority professionals and
corporate employers. These outlets chronicled career opportunities and barriers in "high-tech hot spots" and promoted corporate initiatives such as
3M's internships for students from historically black colleges.[63] *US Black
Engineer's* history reflected the hybridity between mutual aid and corporate
relations. The magazine started in the early 1970s as Cornell University's
Black liberation newspaper *Umoja Sasa* (Swahili for "Unity Now"). In 1978,
entrepreneurial student Tyrone Taborn realized that advertisements from
IBM and other technology corporations, which were seeking Black appli-
cants, could be a lucrative revenue stream. He then struck a deal with the
NSBE to make *Umoja Sasa* its official publication and renamed the magazine
US Black Engineer (the *US* stood ambiguously for Umoja Sasa and the United
States). Taborn then formed the Career Communications Group, launched
Hispanic Engineer, and developed a networking service to connect readers
with corporate human resource offices.[64]

Finally, Richard Florida's theory of the creative class further tightened
the connection between diversity and innovation. *The Rise of the Creative
Class* opened with a Rip Van Winkle tale that identified workforce diver-
sification as among the most dramatic changes in American life since the
1950s. Utilizing statistics such as the Gay Index, Florida argued that the
most talented individuals were drawn to tolerant environments in which
they were empowered to "express themselves and validate their identities."
Florida argued that diversity had become not just a legal necessity, but "a
matter of economic survival" for companies and communities. At the same
time, he wrung his hands about racial disparities in innovative regions.
"Several of my interviewees," he wrote, "noted that a typical high-tech
company 'looks like the United Nations minus the black faces.'"[65]

EMPOWER THE NATION

Sunny affirmations of empowerment and diversity as drivers of innovation took place against the darker background of the Los Angeles riots, rising urban poverty, and a backlash against affirmative action. Managers and consultants debated the impact of this political environment on their internal policies and outward-facing social responsibilities, which shaped their embrace of multiculturalism as an ideal.[66] At the same time, experts scaled up empowerment as a basis for innovations in social policy.

Kanter was again a first mover. In 1986, she worked with Colorado senator Gary Hart on the Democratic party's economic platform. Then, in the 1988 presidential election, she served as Massachusetts Governor Michael Dukakis's economic advisor. In *Creating the Future: The Massachusetts Comeback and Its Promise for America*, the two argued that the state's "economic miracle" resulted from supporting entrepreneurs, fostering regional innovation clusters, building infrastructure, and eliminating red tape. Kanter portrayed Dukakis as a change master who worked with stakeholders to achieve "strength through diversity." Dukakis, for his part, explained his economic platform as a set of "power tools" that spanned venture capital legislation and daycare programs.[67]

Conservatives also developed an alternative empowerment ideal.[68] A key leader was Civil Rights activist Robert Woodson, who was awarded a 1990 MacArthur genius grant for his National Center for Neighborhood Enterprise (NCNE), a grassroots network for Black economic self-sufficiency, crime prevention, and urban renewal.[69] In its Agenda for Black Progress, NCNE contended that the welfare bureaucracy fostered dependency and that "the road from poverty to empowerment [was] one of entrepreneurial risk and exploited opportunities."[70] During the Bush administration, empowerment became a wider buzzword that united "do gooders" and "cost-cutters" in reforms from urban renewal to school choice.[71] Meanwhile, activist Paul Weyrich used digital media to mobilize partisans via a 24/7 interactive satellite network for conservatives that he named "National Empowerment Television (NET)."[72]

During the 1992 presidential campaign, empowerment became a bipartisan neoliberal frame. In *Reinventing Government: How the Entrepreneurial Spirit is Transforming the Public Sector*, consultants David Osborne and Ted Gaebler translated Kanter's arguments into a "third way" for "pushing control out of the bureaucracy, into the community."[73] At the same time, members of the US House of Representatives created a Congressional Empowerment Caucus while The Empowerment Network Foundation (TEN) formed to address urban poverty in the aftermath of the LA riots.[74]

President Clinton then made empowerment a central theme of his domestic agenda. Clinton used the concept both to defend affirmative action and to justify welfare reforms, declaring that the "most important job of our government in this new era is to empower the American people to succeed in the global economy."[75] The centerpiece of his administration's urban renewal efforts was the "empowerment zone," a reformation of Reagan-era "enterprise zones" that aimed to use public–private partnerships to revitalize minority communities in struggling cities.[76] Clinton also created yet another innovation prize—the Ron Brown Award for Corporate Leadership—which recognized "distinctive, innovative, and effective" programs in areas of work/life balance, economic development, environmental management, and workforce diversity.[77]

The Clinton administration also addressed access and equity to computers and the internet through the lens of empowerment. Despite claims that innovation would improve the lives of all Americans, it was clear by the mid-1990s that the benefits of digital technology were unequally distributed. In 1995, the Department of Commerce described economic and racial disparities as a "digital divide" and emphasized that the National Information Infrastructure was committed to "empowering the information disadvantaged."[78]

Reformers adopted the concept of a digital divide to characterize the lack of Black professionals in the innovation economy. The *San Francisco Chronicle* sounded the alarm that Blacks and Latinos were missing out on the technology boom, documenting that only 4 percent of tech company employees were Black and 7 percent Latino, despite representing 8 percent and 14 percent of the workforce, respectively. Using freedom of information requests, the *Chronicle* identified affirmative action violations at most major tech companies, including Apple, Netscape, and Oracle.[79] Critics also noted the divide between the creative enclaves of Palo Alto and Cupertino and the surrounding Black and Brown communities of East Palo Alto and San Jose where assembly line and service workers struggled to make a living.[80] In 1999, Jesse Jackson partnered with Black leaders in Silicon Valley, including Roy L. Clay, Sr., a former HP engineer and founder of Rod L. Electronics to challenge ongoing discrimination in the computer industry. At the Digital Divide and Empowerment Rally, Jackson celebrated "Black innovators [who] have made positive contributions to the computer industry" but admonished that "we have to ask the question: 'why isn't high tech beating a path into place where they can find more such pioneers?'"[81]

TOKEN ENTREPRENEURS, HIDDEN FIGURES, AND CORPORATE ROCKSTARS

Diversifying innovation was as much a task of creating new images of success as it was of structural reform. The very title of Kanter's *Change Masters*, for example, aimed to identify unheralded intrapreneurs. From the 1980s onward, inclusive innovation's proponents invested significant effort to broaden who counted as an innovator.

High-tech boosters' first efforts to demonstrate that innovation culture was equitable were steeped in gendered and racialized tropes. A 1984 *Fortune* photo essay, for example, championed "A Friendly Frontier for Female Pioneers." Its spotlight of nine women founders claimed that the software industry's youth and codevelopment with the women's movement made it "hospitable" to women entrepreneurs compared to the boys' club of microelectronics. *Fortune* quoted Lorraine Mecca, founder of software distributor Micro D, who argued that venture capital was sex-blind and explained "there's no old boy network." *Fortune* also celebrated gender-role reversals. Mecca said of her husband, the company's chairman, "I don't pay him; I give him an allowance. He needs me and I keep him happy." Even as the magazine celebrated equal opportunity it followed sexist patterns, for example, identifying motherly accounting software founder Stina Hans (photographed hugging her young son) and militant Software Corps of America president Shirley Eis (staring severely at the camera). *Fortune* captured Mecca, a "water ballet enthusiast," in a bathing suit.[82]

Tech journalists in the 1980s adapted existing innovator personas to women entrepreneurs in tokenistic profiles. ASK Computer Systems founder Sandra Kurtzig, for example, was cast as the female Steve Jobs. The *San Jose Mercury News* recounted how she had developed her program MaMa (short for "Manufacturing Management") from the comforts of her apartment. She then built her software, which streamlined purchasing and production for large companies, into a multimillion-dollar business. Kurtzig "looked like a well-kept, polished woman on her way to a country club luncheon," but she "crackled with the cocky energy of a jock." Like other Silicon Valley pioneers, she worked hard, putting in twenty-hour days, and played hard, showing up to a business lunch on Halloween in a satin gown as a pregnant bride in a shotgun wedding. Kurtzig, for her part, embraced Silicon Valley's libertarian ethic and balked at those who called her the "epitome of feminism."[83]

Representations of women innovators improved with the rise of corporate feminism. In 1989, *Business Week* spotlighted ten women "scaling high-tech heights" without mentioning their marital or maternal status. It praised their successes but also explored industrywide challenges of tokens, wage gaps, sex discrimination, and the need for structural changes to attract more women into technoscience. *Business Week* also emphasized the role of intrapreneurship in launching entrepreneurs such as Adele Goldberg who had worked on Xerox PARC's pioneering graphical–user interface before leaving to create her own company.[84]

Profiles of Black and Brown innovators in mainstream media were comparatively absent.[85] *Black Enterprise* and *US Black Engineer* filled the void by spotlighting "young gifted black people" who had "infiltrated government, industry, and public service in unprecedented numbers."[86] These publications created annual awards to give minority innovators the "honor and recognition they truly deserve."[87] Their narratives of hiddenness followed the pattern of Kanter's change masters, reflecting the reality that most Black scientists, engineers, and businessmen worked in large organizations. A common theme was the recognition that to achieve greater success, Black innovators needed to start their own ventures. *Black Enterprise*, for example, profiled AT&T manager turned independent consultant Lou Burke whose journey echoed Briggs's at Chipco. Before branching out on his own, Burke "spent much of his time" convincing "top management to change its outdated way of thinking," and through his initiative, AT&T implemented "decentralization and flexibility to promote innovation throughout the organization."[88]

The internet boom shifted these narratives of minority innovation from middle managers to "Black digerati" who were "masters and creators of the Digital Revolution."[89] Media profiles explored the contradictions of digital invisibility. On the one hand, *Business Week* highlighted entrepreneurs who were "invisible—and loving it" because internet commerce eliminated many outward signs of race that led to discrimination.[90] On the other, *US Black Engineer* lamented that Mark E. Dean, inventor of IBM's personal computer architecture, was "High-tech's 'Invisible Man.'"[91] Advocates worked to elevate Silicon Valley pioneers, including Roy L. Clay, Sr. and Marc Hannah, in a similar fashion to those used by activists in the 1970s.[92]

The most prominent representations of women and minority innovators were neither token entrepreneurs nor hidden race heroes; they were corporations. Tech industry recruitment advertisements promoted their cultures of "diverse and multi-talented professionals."[93] Johnson & Johnson used Black History Month to seek future "Pioneers. Innovators. Trailblazers."[94] Microchip firm Advanced Micro Devices declared that its multiculturalism was "a springboard for innovative, insightful thinking . . . We don't all see alike. We don't all talk alike. We don't all feel alike. We don't all think alike. But we all like results."[95] The Japanese electronics conglomerate Sony linked its "spirit of innovation" with Black achievement in media arts. Its Sony Innovators Program provided awards to talented and transgressive people of color in an initiative that enlisted Quincy Jones and Herbie Hancock as "master innovators."[96]

Corporations used the fusion of empowerment, diversity, and innovation to define their brands as progressive and future-oriented. Apple's famous "Think Different" campaign associated the company with diverse pioneers that included Amelia Earhart, Mahatma Gandhi, Martha Graham, and Martin Luther King Jr. In contrast, Intel's "Sponsors of Tomorrow" spotlighted the company's multicultural intrapreneurs. One memorable ad depicted USB coinventor Ajay Bhatt, a middle-aged, sweater-vest-clad Indian engineer walking in slow motion through throngs of swooning coworkers to the sound of electric guitars. "Our rockstars," Intel boasted, "aren't like your rockstars." Microsoft took a patriotic approach that could have been written by Kanter herself. The company defined Windows 98 as "an important step forward on the path of innovation that has fueled an era of opportunity, empowerment, and unprecedented economic growth for America."[97]

EMPOWER EVERYTHING

By the early 2000s, empowerment was synonymous with the diversity and tolerance at the heart of the creative class. Innovation experts championed diversity as a key driver of new ideas and products.[98] Nearly every corporate mission statement emphasized the themes. The rhetoric was especially prevalent in digital platform companies; for example, Facebook's mission was to "give people the power to share and make the world more open and connected."[99]

At the same time, disparities in participation and high-profile discrimination cases received consistent media attention as Silicon Valley extended its global and cultural power. A 2016 policy study of STEM innovators, for example, found that just 12 percent of innovators were women and only 0.5 percent of US-born innovators were Black.[100] Another study revealed that Asian Americans, and Asian American women in particular, lagged behind white men and women in senior leadership.[101] Meanwhile, magazines such as *The Atlantic* asked, "Why Is Silicon Valley So Awful for Women?" and the survey "Elephant in the Valley" found shocking levels of tech industry discrimination.

In continued efforts to address inequity and unlock the benefits of diversity, dozens of initiatives contributed to improving conditions for women, Black, LGBTQ, and other minoritized innovators.[102] Organizations such as the National Center for Women and Information Technology (NCWIT) collaborated with Google and other large corporations on corporate training, demographic data sharing, and positive portrayals of women in tech.[103]

Some feminist and antiracist critics came to attack the very identity of "innovator" as a barrier to inclusion. In the 2010s, the epithet "tech bro" emerged as a stereotype for the innovation economy's "toxic masculinity." Tony Ruth's satirical art project *Businesstown* linked this negative persona to the hidden values of a corporate ecosystem that gave all the glory to entrepreneurs, but was run by less glamourous forms of labor. While "Elon Muskrat" designed his hover train, "Brogrammers" took advantage of their companies' quality-of-work life programs to "crush code" and "get some reps in before lunch."

Tensions between empowerment as a source of equity and as an apologia for corporate capitalism eventually sparked an open battle in innovation culture. At the center was media billionaire Sheryl Sandberg's *Lean In*, which sold over two million copies and used Sandberg's life story as an argument that workplace equality and the imperative to innovate were mutually constitutive. She accompanied her message with a self-help program "to empower women to achieve their ambitions" through a worldwide parallel organization of "lean-in circles."[104] *Lean In* also sparked backlash among critics who attacked the "oxymoron" of corporate feminism. Feminist icon bell hooks was especially harsh, arguing that Sandberg's "feminist rhetoric"

The Brogrammers
crush code then bail to get some reps
in before lunch.

dubstep

kettlebells

5-hour energy

FIGURE 7.2

The Brogrammers of *Businesstown*. Critics in the 2010s pointed out the conformism and "toxic masculinity" of innovation culture with the epithet of "tech bro." Courtesy of Tony Ruth.

was "a front to cover her commitment to western cultural imperialism, to white supremacist capitalist patriarchy."[105]

The ease with which visions of empowerment and difference spanned feminist critique and innovation evangelism reveals a set of affinities among the two: a shared belief that pragmatists working in collaboration can replace entrenched power structures; that social change and technological change are mutually reinforcing; that individualism and communitarianism are compatible; that individual, organizational, and societal goals are in alignment; and that we can all participate in innovation and all be empowered by it.

The Innovation Strategist works with clients
to define problems, ideate solutions, and
articulate outcomes.

FIGURE 7.3

The Innovation Strategist. This satirical profile captures the tensions between ambition, feminist theory, and corporate service of innovation expertise. Courtesy of Tony Ruth.

Rosabeth Moss Kanter, the first inclusive innovation expert, embodied this hybridity and its contradictions. Her career showcases the strengths and limitations of empowerment as a reformist strategy. Today's critics might dismiss her for making empowerment a corporate tool of innovation culture, much as Sandberg became a target. However, Kanter's concepts remain an essential foundation of alternatives to supremacy and patriarchy. For example, the radical labor collective FemTechNet—which aims to use "civil rights, anti-racist, queer, decolonizing, trans-feminist pedagogies"

for the "radical redistribution, reinvention, and repurposing of technological, material, emotional, academic, and monetary resources"—defines itself
succinctly as a *power tool*.[106]

Still, *Businesstown* provides a disconcertingly accurate proxy for the real
world. Few women and minority innovators have found a place in its ecosystem. Among the dozens of social types profiled, only the company bus
driver was coded Black and only three denizens were female. There was the
Corporate Fertility Consultant (the mother) who "offers advanced child-
creation solutions which empower employees to devote their best years to
company goals unhindered by family distractions"; the Social Responsibility Specialist (the militant) with triple-bottom-line tools; and the Innovation Strategist, whose messages came with a certain seductiveness, and who
aspired to be more than the corporation's pet.

8 LIFELONG KINDERGARTEN

You can't be a serious innovator unless you are willing and able to play.
—Michael Schrage, *Serious Play*, 2000

Why did James McLurkin build his army of robotic ants? Maybe the young Black "troublemaker" wanted payback for the teacher that sent him to the corner because he couldn't sit still. Perhaps he had something to prove to his father, who told him to stop "fooling around" with toys. There was also the enabling cousin on whose computer a young McLurkin learned to program video games. Whatever the case, by the time McLurkin was fifteen, he had turned a discarded remote-control car into a working robot. Later, at MIT, he gained access to all the tools a troublemaker could want. One fateful weekend, he snuck his friends into the Artificial Intelligence Lab to assemble his swarm of ant-sized microrobots. The ants promised to revolutionize surgery, avert terrorist attacks, and remediate environmental disasters.[1]

For experts at the Smithsonian Institution's new Lemelson Center for the Study of Invention and Innovation, understanding the roots of McLurkin's innovation was crucial to America's twenty-first-century success. His was one of a handful of "innovative lives" displayed at the National Museum of American History (NMAH) for its 2002 exhibition *Invention at Play*. McLurkin shared the best traits of past inventors like Alexander Graham Bell, but he applied them to "lift up" minorities and to "make things for the good of society."[2] He was also living proof that while innovation

FIGURE 8.1
James McLurkin demonstrates his microrobots as an example of playful invention.
Photo by Donna Coveney. Courtesy of MIT Media Office.

required infrastructure, research, money, and education, you first needed
to mess around with toys.

Identifying the sources of new ideas was only one step in the Lemelson
Center's plans to inspire young innovators. *Invention at Play* gave visitors
a hands-on opportunity to unlock their inner McLurkin. In a partnership
with the MIT Media Lab, museumgoers could become musicians by stretch-
ing Play-Doh to make electronic sounds. Using microcomputers called
"crickets," children imitated different instruments via colorful block code
that was inspired by the improvisation of jazz musician Thelonious Monk.[3]
Nearby, kids built drawing robots that spun markers in chaotic circles.

Others laughed as their parents took turns riding a mockup sailboard, the result of a lifetime of incremental improvements on an initial bold idea by inventor Newman Darby. Every story and every activity in the exhibition underscored the same idea—creative play was essential to innovation.

This chapter explores how the imperative to innovate came to encompass the youngest Americans. I investigate how from the 1980s to the early 2000s, a new coalition of experts trained in education, psychology, human development, and museum design identified children as future innovators. I then explore the role of philanthropic foundations in shaping childhood innovation programs, many with special emphasis on reaching girls and underrepresented minority children. In these initiatives, educators seeking to enhance childhood creativity made common cause with policymakers concerned with national competitiveness and technology companies seeking new markets. This coalition made *play* a virtue for innovators of all life stages.

"WHERE ARE TOMORROW'S INNOVATORS?"

Until the 1980s, an emphasis on children as potential innovators was rare in the United States. Because innovation training initiatives focused on technology development and economic growth, their youngest targets were typically university students. The competitiveness crisis, however, laid the groundwork for a national movement to harness children for future roles in the innovation economy. The aim of these programs was to capture children's attention before they became conformists and their potential was lost.

The search for young innovators was the latest iteration in longstanding debates about the place of creativity in human development. Since Friedrich Froebel opened his first kindergarten in 1837, progressive educators have argued that young people have a natural capacity to create that can be cultivated or suppressed depending on the environment in which they were raised.[4] Throughout the twentieth century, educators from Maria Montessori to John Dewey, Jean Piaget, and Lev Vygotsky developed pedagogy with the conviction that children were developmental learners who constructed their self-understanding and the foundations for democracy through play. This exploratory philosophy was in tension with mass public education and its demands for standardization and accountability.[5]

Cold War education reforms, however, linked creativity to national survival. After Sputnik, the US federal government invested hundreds of millions of dollars in curricula, teacher training, and after-school programs such as science fairs to build a workforce that could compete with the Soviet Union. These initiatives emphasized national standards, scientific rigor, and the "pipeline" of scientific workers. Demands for manpower and rigor nonetheless were entangled with the celebration of creativity, the rejection of conformity, and the cultivation of democratic thinking.[6] In this milieu, intelligence testing served as a means of establishing creativity as limited to the lucky few identified as "gifted and talented."

During the competitiveness crisis, policymakers began to explore public education as a tool for closing the innovation deficit. During the last days of the Carter administration, the Department of Commerce hired the educational consultant Anne Branscomb, wife of former National Bureau of Standards director Lewis Branscomb, to survey the state of creativity in American schools. Her 1980 report, *Learning Environments for Innovation*, emphasized that "participation in innovation" was a team endeavor that should not "be limited to an elite few" but rather depended on skills that American schools could cultivate. However, the report identified cultural and institutional forces that discouraged creativity among Americans, including representations of innovators as male loners, a lack of opportunity for women and minorities, cuts to school budgets, too much testing, and the neglect of gifted students. To support every child's inner innovator, Branscomb recommended investments in informal learning that included using toys as teaching tools, building hands-on science museums, and creating educational television programs.[7]

After Carter's defeat, Branscomb's report gathered dust, but other experts built on its themes from the top down and the bottom up. At the federal level, the Reagan administration made the failures of American public schools a key theme of domestic policy. The 1983 manifesto *A Nation at Risk* emphasized the need for rigor and common standards to address "mediocrity" in the nation's education system.[8] But *A Nation at Risk* did not address innovation education as a goal. Instead, the blueprint for such instruction again came from the Department of Commerce.

In 1988, Donald J. Quigg, the commissioner of the U.S. Patent and Trademark Office, launched the Inventive Thinking Curriculum Project, or Project XL for short. Quigg was neither an educator nor a career policymaker;

he had spent the prior thirty-five years as a patent lawyer for Phillips Petroleum in the oil towns of Oklahoma. He had two goals for Project XL: to prepare children to be innovators and to revive "the spirit of those golden years . . . when inventors were heralded as true American heroes."[9]

Project XL was unsubtle in its deficit alarmism. "Imagine your office staffed with underskilled, undereducated workers, and your clients all from foreign shores," its materials warned while citing a dramatic increase in patents issued to "foreigners." Asking "where are tomorrow's innovators?" the initiative identified children as the nation's "most precious natural resource." To extract that resource, Commerce developed and distributed materials to 80,000 American classrooms via lesson plans, extracurricular contests and events, and a partnership with the children's magazine *Weekly Reader*.[10]

Project XL channeled grassroots invention education among teachers, educational researchers, and civic organizations. Quigg assembled a team of teachers and educational experts to develop and carry out the initiative. A key contributor was Buffalo, NY second-grade teacher Marion Cañedo. In 1979, a decade prior to the Commerce program, Cañedo had asked her students to "find some small but meaningful thing that needed doing—and then invent a way to do it." Her students generated over two hundred inventions, which they displayed throughout the school. Cañedo began connecting with other teachers across the country who were independently developing similar initiatives.[11] Quigg got word of Cañedo's successes and, in 1987, invited her to the Department of Commerce to lead *Invent America!* the nation's first "invention convention." *Invent America!* further linked these regional extracurricular events in a new National Inventive Thinking Association.

For Project XL's curriculum, Quigg tapped leading creativity theorists, including Donald Treffinger, an educational psychologist at Buffalo State University's Center for Studies in Creativity, and Calvin Taylor, a critical thinking advocate and head of the Alabama-based teacher consulting firm Talents Unlimited. Their "inventive thinking" curriculum emulated design methods being developed concurrently for college students at MIT, Stanford, and CMU. Lesson plans taught children how to identify problems relevant to their lives, brainstorm solutions, build designs and models, develop patents, and even create advertising jingles.

In pursuit of a culture of "excellence," Project XL linked invention to the American past and to its diverse, prosperous future.[12] The curriculum instructed that Alexander Graham Bell and Thomas Edison had begun their inventive streaks as children. History was also a strategy for encouraging more girls and minorities to pursue invention. Project XL collaborated with the National Technical Association and the NAACP's Science Olympics competition to create middle- and high-school curricula on Black Innovators in Technology that profiled past inventors such as Charles Drew, George Washington Carver, Garrett Morgan, and Mildred Smith, an initiative that was aided by a series of new children's books about Black innovators.[13]

A variety of local, regional, and national childhood invention and innovation programs flourished alongside Project XL.[14] These initiatives offered a broad tent for classroom teachers, university experts, policymakers, corporate sponsors, and nonprofits. Their targets included suburban gifted and talented programs as well as urban and rural "at risk" schools. Some predated Project XL, such as the IBM-funded Odyssey of the Mind competition, started by a New Jersey industrial design professor in 1978. Others emerged in tandem. For example, in 1989, the Academy of Applied Sciences, a New Hampshire–based, women-run nonprofit established by four teachers, created the Young Inventors Program to broaden participation in innovation. Its *I Was Meant to Invent* handbook profiled children and women inventors from American history to reinforce values of curiosity, tinkering, persistence, and sharing.[15] Also in 1989, independent inventor Dean Kamen established the FIRST (For Inspiration and Recognition of Science and Technology) robotics competition in collaboration with MIT engineer and designer Woodie Flowers. Funded by Autodesk and the American Society of Mechanical Engineers, FIRST deliberately emulated athletic competitions, with masculine team names such as the Bomb Squad and the Warriors, to show "young people that engineering and science [could] be as captivating and exciting as a sporting event."[16]

It is difficult to gauge the full impact of invention contests and conventions on a generation of American children. However, if promotional materials and news reports were to be believed, the kids were having a blast. *The Dallas Morning News*, for example, interviewed prize-winning sixth-grader Deanna Whitmore, inventor of the "Super Brush" that had a built-in hairspray dispenser. An AP story meanwhile showcased second-grader Kristin Giallella, creator of a longnecked coat hanger for short people. According to the nine-year-old, her peers "didn't really think it would do any good," but she proved them wrong.[17]

As invention conventions flourished, an important parallel development in childhood innovation took shape at the intersection of computing
and psychology. In the 1970s, MIT professor Seymour Papert became convinced that, in the hands of children, computers could be powerful tools to
remake society. Papert had been trained as a mathematician, but his interest in how people acquire mathematical knowledge drew him to cognitive
science and psychology. In 1963, after studying as a protege of Piaget, Papert joined MIT to codirect its new Artificial Intelligence Laboratory. At the
time, computers were rare, expensive, and difficult to use. Moreover, most
artificial intelligence researchers sought to mimic or surpass human brain
power. Papert, in contrast, emphasized that computers were tools for augmenting human thought. Throughout the 1970s, he codeveloped LOGO, a
programming language for children. In his 1980 book *Mindstorms: Children,
Computing, and Powerful Ideas*, Papert argued that the computer revolution
had the potential for "more personal, less alienating relationships with
knowledge" and that LOGO was the first step. Disdainful of the factory
model underlying the educational system, Papert looked to Brazilian samba
schools in which dancers and musicians learned together through "cohe-
sion," "belonging," and "common purpose."[18]

Papert and his students and colleagues extended this work from the
laboratory into the world with support from the NSF, the MacArthur Foundation, IBM, Apple, and LEGO. In Project Headlight, fourth-graders in an
inner-city Boston school with a majority Black and Hispanic population
used LOGO to create their own programs for solving fractions. The researchers argued that students taught in this constructivist method performed
better at math, programming, and problem-solving skills than control
groups taught in subject-based silos.[19] The LOGO programming language
was also adopted across the country as elementary schools built dedicated
computer labs.

In 1984, Papert became a founding member of the MIT Media Lab. He
and his colleagues established the innovation center as a global leader in
research into children and technology. MIT professor Jeanne Bamberger,
for example, the director of the Laboratory for Making Things, collaborated
with local public schools to foster "active makers and builders of knowledge," rather than "passive consumers."[20]

In addition to its pedagogical benefits, there was money to be made in
childhood computing. In the early 1980s, two of Papert's students, Mitch

Resnick and Stephen Ocko, collaborated to link children's software with exploratory learning tools in the physical world. Resnick was a former *Business Week* reporter who had covered Silicon Valley during the competitiveness crisis. Ocko had experience as an architectural model-maker, documentary filmmaker, and entrepreneur. Their medium for bridging the material and the digital worlds was the ubiquitous LEGO brick. While the MIT graduate students were developing their project, LEGO contacted Papert to learn more about his laboratory. Papert and LEGO struck a deal that made the toy company the lab's primary sponsor, which led to LEGO's computer-connected Mindstorms product line for use by schools and adult hobbyists.[21]

Resnick stayed at the MIT Media Lab after earning his PhD, where he established the Lifelong Kindergarten Group. He and colleagues emphasized the power of belonging and collaboration in young people's learning and social development. Group member Natalie Rusk, for example, established the Computer Club House national network in 1993 to provide inner-city kids access to technology on their terms.[22] Resnick believed that the Lifelong Kindergarten Group's methods could apply beyond young children and that "the rest of school (indeed, the rest of life) should become more like kindergarten."[23]

By the late 1980s, in short, there was a multifaceted movement for childhood invention and innovation in the United States. Its advocates worked at the interface between public education, government, academia, and industry. They imagined children to be at once a national resource for economic prosperity and the source of radical creative futures. In the 1990s, they were joined by deep-pocketed brokers.

FOUNDATIONS AND LIFE STAGES

Since the Gilded Age, philanthropic foundations created by industrial titans, including the Carnegie Corporation, Rockefeller Foundation, and Ford Foundation, have been important behind-the-scenes players in American culture.[24] These charitable organizations are powerful mediators that perform their work in indirect, but important ways. First, and most obvious, their donations shape which problems get addressed. Foundations are also brokers of partnerships that build wide-ranging networks around their issues of choice. Their influence helps shape whose ideas matter and whose are stigmatized.[25] Finally, to carry out their missions, foundations have served as important job creators and training grounds for innovation experts.

In the 1980s, federal policy spurred experimentation and growth in the American philanthropic community. As part of his ideology of limited government, President Reagan established a Taskforce on Private Sector Initiatives that aimed to privatize solutions to public problems. With the motto "Building Partnerships USA," the taskforce joined in an uneasy alliance Reagan officials seeking to reduce social welfare programs with the leaders of private foundations who wanted to improve those same social programs through better coordination of state and private resources.[26]

Meanwhile, a new set of foundations specializing in innovation emerged alongside established giants like Carnegie and Ford. In Silicon Valley, the combination of rapid growth and widening inequality had produced an uneasy mix of civic pride, altruism, hubris, guilt, and fear of regulation. Regional foundations invested in cultural and educational initiatives using the methods that had made their donors successful in business. The Bay Area Entrepreneurs Foundation, for example, used equity stock to invest via "venture philanthropy" techniques to support education and youth initiatives that included a collaboration with the nonprofit Partners in School Innovation.[27] In Kansas City, pharmaceutical magnate Ewing Marion Kauffman shifted his educational giving from drug abstinence to entrepreneurship education.[28] Collectively, these programs distributed hundreds of millions of dollars to make innovators.

In 1992, a new philanthropy, The Lemelson Foundation, set out to foster innovators at every stage of American life. In pursuit of this goal, the foundation established connections between a vast network of government agencies, universities, NGOs, associations, and primary and secondary schools. The foundation's benefactor was its namesake, the independent inventor Jerome Lemelson.

Lemelson was a controversial figure in innovation culture. His more than six hundred patents put him on the short list of most prolific inventors in American history. His work covered everything from toys to robots, cancer detection devices, and fax machines. But Lemelson made his fortune through litigation and licensing rather than entrepreneurship and product development. Born in 1923, Lemelson grew up on Staten Island where he trained as an aeronautical and industrial engineer. After a brief stint in the world of corporate engineering, he left with the ambition of becoming an independent inventor. Lemelson filed his first patent in 1953, for a propeller beanie that a child could make spin by blowing into a tube. But after experiencing trouble selling his invention, Patent Office bureaucracy, and experiences of having his ideas stolen, he turned to the legal system.[29]

Lemelson anticipated the technical directions of different fields, imagined future devices that matched those directions, earned patents, and then either demanded licensing royalties or sued corporations for infringement when they capitalized on the direction. By the early 1990s, he was thought to have made over a billion dollars with this strategy from major US and Japanese automakers and electronics firms. Fellow independent inventors lauded him as a Robin Hood for his "true grit" in taking on greedy, innovation-stifling corporations. Technology companies, in contrast, accused Lemelson of "patent blackmail," which they claimed stifled innovation by using "submarine" patents that made it more difficult to bring innovative products to market.[30]

As Lemelson's wealth skyrocketed, he established his foundation to champion invention and to venture into policymaking. In 1992, the inventor announced his Lemelson National Program. Drawing on the expertise of MIT professor Lester Thurow, Lemelson echoed the message of Project XL, arguing that America had lost its "productive edge."[31] The United States was falling behind Japan and other industrialized nations, and students were pursuing entertainment and sports careers instead of careers in invention and innovation. The premise of his National Program was that with training in scientific invention and entrepreneurship, the next generation of innovators could reverse the trend and lead to thousands of new companies and jobs.

The Lemelson Foundation used a pyramid model to cultivate innovators across the entire human life cycle.[32] At the top, the foundation established a partnership with MIT to honor innovators. The MIT prizes included a $500,000 award for a practicing American innovator that resembled Nixon's award model, as well as a lifetime achievement prize and a graduate student prize. The first cash award went to Saturn GM engineer William J. Bolander with subsequent winners including appropriate technologist Amy B. Smith, very-large-scale integration circuit designer Carver Mead, and drug designer Robert S. Langer. Lifetime achievement recipients included Jacob Rabinow and Douglas Engelbart, Genentech founders Herbert Boyer and Stanley

FIGURE 8.2

The Lemelson National Program announced its mission in major newspapers throughout the United States contrasting the power of invention and innovation with failing political and cultural systems.

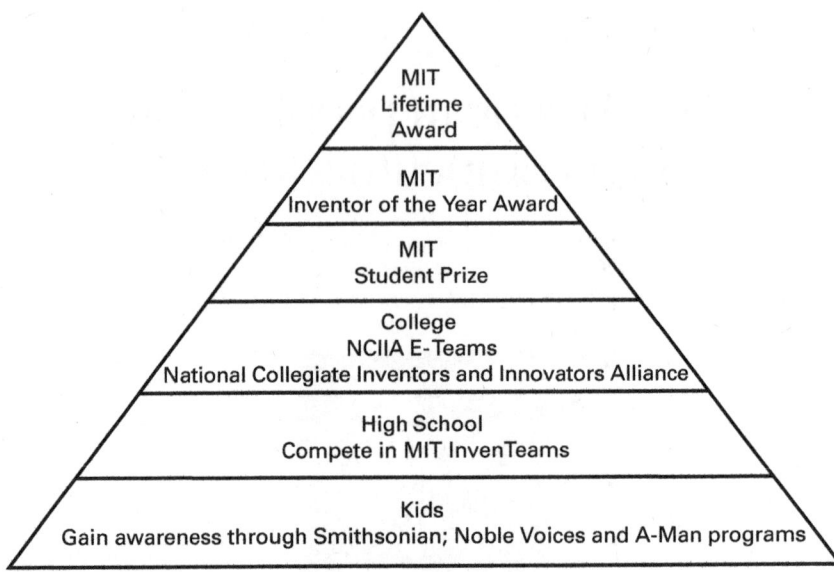

FIGURE 8.3
The Lemelson Pyramid. The Lemelson Foundation aimed to cultivate innovation at
all life stages with age-appropriate educational programs and incentives.

Cohen, Kevlar inventor Stephanie Kwolek, and pharmaceutical innovator Gertrude Elion. Graduate awards targeted MIT students, including the
robot ant maker McLurkin.

Moving down the pyramid, the Lemelson Foundation funded the
National Collegiate Inventors and Innovators Alliance (NCIIA). This undergraduate network built on pilot programs at MIT and Hampshire College,
a small liberal arts school in Amherst, Massachusetts, where the foundation established a disability-oriented Center for Assistive Development. The
NCIIA was a deliberate riff on the National Collegiate Athletics Association
(NCAA), sponsoring an invention competition—akin to the NCAA's March
Madness basketball tournament—in which student "E-Teams" practiced
the basics of venture formation. The "E" stood for *excellence* and *entrepreneurship*.[33] The NCIIA grew rapidly to include over one hundred member
institutions from research powerhouses like MIT to regional colleges such
as Nebraska Wesleyan University.[34] As the NCIIA expanded, it became the
independent nonprofit VentureWell.[35] Similar to MIT's Innovation Center
in the 1970s, student projects clustered around innovations in sports, medical devices, and environmental technologies.[36]

The Lemelson Foundation distinguished itself from predecessors and peers at the base of the pyramid. In the 1980s and 1990s, innovation experts had experimented with how best to reach American kids, the largest potential source of innovative talent but with the least immediate return on investment. Should initiatives focus on attempts to reform public schools through teacher education and curriculum design? Should they instead create after-school gifted and talented programs in wealthy districts and computer clubhouses in underserved urban communities? Through invention conventions and robotics competitions? Via television and edutainment software? Each of these approaches had its strengths, weaknesses, proponents, and detractors.

THE FUTURE IS HANDS-ON

The Lemelson Foundation instead chose to focus its investments in a history museum. In 1995, the Lemelson Foundation presented the Smithsonian's NMAH with $10 million to endow the Lemelson Center for the Study of Invention and Innovation. At the time, it was the largest cash gift from an individual in the Smithsonian's history.[37]

Why did The Lemelson Foundation select a history museum as their preferred site to train American kids to be innovators? Like the philanthropic foundations that fund them, museums have less direct influence over patrons than schools, but they reach a large audience with less regulation and at lower costs. Museums serve as meeting grounds for people and organizations from multiple walks of life. Despite their stodgy reputation, museums also can be incubators of new ideas that blur the hierarchies of knowledge found in schools and universities. Museums were especially well positioned to promote innovation to children as a distinctly American quality, particularly among children who might not otherwise view themselves as potential innovators.

Lemelson's Smithsonian endowment was part of a worldwide trend to harness museums and science centers as sites of innovation education. Efforts elsewhere included the rebranding of the National Inventors Hall of Fame in Akron, Ohio, as Inventure Place and the installation of a major new Innovation Station at Dearborn, Michigan's, Henry Ford Museum.[38] Museums also put special emphasis on past and present minority innovators, including the Smithsonian Anacostia Museum's *Real McCoy* exhibition and the Chicago Museum of Science and Industry's traveling exhibit Black Achievers in Science.[39] Collectively these initiatives reached millions of children through individual visits, class field trips, and invention conventions.

The largest of the new childhood innovation centers was San Jose's Tech Museum of Innovation. The Tech was the result of twenty years of philanthropic planning that began when the Junior League of Palo Alto, a group of Silicon Valley founders' wives led by Lucile Packard, proposed a learning center to "make technology accessible" to residents. In 1984, San Jose won a competition to host the center as part of a larger downtown urban redevelopment initiative. The project however became mired in fundraising challenges and a multimillion-dollar lawsuit as San Jose intended to displace one hundred residents for the project.[40] In the interim, The Tech planners opened The Garage, a temporary innovation education center.

Despite the rocky start, The Tech became a focal point of Silicon Valley philanthropy and educational research. Its board of directors included William Hewlett, Robert Noyce, Steve Wozniak, and Douglas Engelbart as well as educators, civic leaders, university administrators, journalists, and politicians. The Tech had over one hundred corporate sponsors from giants like Apple and Lockheed to video game and biotech start-ups as well as hundreds of individual Valley residents. In 1987, The Tech spearheaded Project Mindstorm to bring Papert's computer education from Boston to Silicon Valley. It later created a teacher-training institute with collaborators from IDEO, Genentech, Lockheed, NASA, and the California Department of Justice to develop lessons on consumer advocacy, the ethics of innovation, and straw bridge–building contests.[41] When The Tech finally opened in 1998, it partnered with the Department of Commerce on an inaugural exhibit on "The Spirit of American Innovation" that included events with Black Silicon Valley pioneers, such as Roy L. Clay, Sr., and emphasized "the innovator in everyone."[42]

As a government-sanctioned space for innovation education, the Lemelson Center faced more complex challenges than its private peers. Using history to explain the future was the easy part. The Center was in many ways a call-back to the NMAH's original mission, which opened in 1964 amid the height of the nation's technological optimism as the Museum of History and Technology. The infusion of private philanthropy was also welcome amid tough political and financial times. While the Center was being formed, the nearby National Air and Space Museum was embroiled in controversy because a planned exhibit portrayed the Enola Gay, the plane that dropped the atom bomb on Hiroshima, in a critical light. Morale was low throughout the Smithsonian, but especially at the NMAH, which had outdated exhibits and a disillusioned staff. Nonetheless, the Smithsonian was a federal institution funded by Congress, and its staff consisted of government employees who were deeply skeptical of corporate encroachment into the museum's mission; especially when those resources would be singularly focused on the message of innovation.

The Lemelson partnership was the result of an intrapreneur inside the NMAH. Curator and science historian Arthur Molella was well into the planning for a major new exhibit called "Science in American Life." One of the centerpiece case studies of his exhibit was to be the Superconducting Super Collider (SSC), a multibillion-dollar project at the cutting edge of physics research. In the middle of exhibition planning, however, Congress eliminated the SSC.[43] Meanwhile, Molella's highly successful pilot for the exhibit, the Hands On Science Center, also faced the chopping block. The exhibit combined history with practical experiments to encourage STEM careers. But its primary donor, the American Chemical Society, was backing out.[44] At a fundraising dinner, Molella had a chance encounter with Lemelson, whom he convinced to donate $30,000 by appealing to the inventor's patents on lasers, one of the technologies utilized in the exhibit.[45] Months later, Lemelson's new foundation approached Molella with an offer that was three hundred times as generous.

At first, neither the Smithsonian nor the Lemelson Foundation had a handle on distinctions between invention and innovation, or how to present them to a wider public.[46] This confusion was evident in early programming. The Lemelson Center hosted academic symposia with innovation experts that included geographer AnnaLee Saxenian. Coinciding with the opening of the public internet, the Center turned existing projects on the history of weaving and the physics of quartz watches into interactive online modules. The Center then moved afield from science into art and technology with an exhibit on the electric guitar.

The Lemelson Center found its direction with *Innovative Lives*, a series of profiles that also showcased Lemelson–MIT award winners, including McLurkin and the Kevlar inventor Stephanie Kwolek. The series included gatherings where children met with innovators to learn the sources of their ideas as well as how they overcame the extra barriers faced by women and minority inventors.[47] This was part of a coordinated Lemelson Foundation effort to have its grantees work together to amplify its mission. To that end, the Lemelson Center hosted NCIIA competitions displayed student inventions (such as a snowboard for people with disabilities), and forged partnerships with children's organizations such as A-MAN (African-American Male Achievers Network, Inc.).[48] These gatherings were augmented with programming for use in libraries and classrooms such as the videos *She's Got It!* and *Lewis Latimer: Renaissance Man, African American Inventor*. The Center determined, as had so many experts before them, that they could best contribute to the nation's progress by training innovators.

INVENTION AT PLAY

In 1999, the Lemelson Center hit upon the concept that established it as a national leader in childhood innovation. The Smithsonian's donor agreement required the museum to produce a major traveling exhibit about invention and innovation. During the two decades prior, there had been nearly twenty-five exhibits elsewhere in the country that rehashed the golden age of American invention. The Center was moving along a similar path, but staff worried that existing approaches did not capture their intentions. Interviews with museumgoers confirmed their doubts. Most visitors associated invention and innovation with exceptional individuals and particularly with famous white men from America's past. "The 'inventor' label," they discovered, "does not appear desirable or accessible."[49]

Well into planning, the Lemelson Center team coalesced on *play* as an organizing theme.[50] Almost immediately, staff realized that they had tapped into an innovation of their own. Play had been a concern for creativity experts in the 1950s and influenced museum design in the 1970s, notably at San Francisco's Exploratorium and the child-centered Brooklyn Museum.[51] But in the 1990s, play studies were experiencing a renaissance. Play was a core theme of Papert's work on children's computing, a connection enhanced by the growing market for video games and edutainment. Psychologists had also begun to link play to innovation.[52] In 1995, interdisciplinary scholars Robert Root-Bernstein, Maurine Bernstein, and Helen Garnier provided empirical evidence for the relationship between play and innovativeness using a remarkable data set of interviews with scientists (including four future Nobel Prize winners) taken at intervals from 1958 to 1988. Their research team found that many of the prominent scientists interviewed were also expert musicians, painters, and toy-builders from a young age. This multifaceted creativity was not simply the outcome of personal genius, it had been *taught* to them. Unfortunately, the psychologists concluded, most students received science education through "the watered down pap of curricula and textbooks" instead of the creative way that scientists actually learned.[53] In *The Scientist in the Crib*, another interdisciplinary team, led by psychologist Alison Gopnik, extended these ideas about creative talent to argue that play was an essential activity that helped children develop physical connections in their brain.[54] Invention and discovery resembled the arts, it was *fun*, and it could be taught.

The Lemelson Center recognized that it could translate the work of play experts for a broad public. In 2000, the Center held a two-day symposium, "The Playful Mind," that included the Root-Bernsteins and the MIT Media Lab's Jeanne Bamberger as well as a toy-making workshop where kids could make kinetic sculptures and a session where they listened to improvisational jazz. The symposium provided the groundwork for the Lemelson Center's first signature exhibition, *Invention at Play*.

Invention at Play linked children's engagement in the world with the future practices of successful scientists, engineers, and inventors. To make its case, the Center emphasized the ideas of the Root-Bernsteins in particular, who served as project consultants. Exhibit materials argued that play was universal: "something that all of us, especially children, engage in naturally, wholeheartedly, and as often as possible." Play, moreover, had no fixed goals, it "is engaged in for its own sake" because it "is open-ended and absorbing . . . deeply satisfying." But play was not all fun, it could be "arduous, frightening and time-consuming." This "frustrating play," however, was valuable because it cultivated individual perseverance and collaborative problem-solving. These values and experience of play were building blocks of the *mindset* required for invention and innovation, such as "curiosity, persistence, imagination, communication, problem solving," and the *skillset* for "manipulating and understanding the properties of the material world." The exhibit's argument was deliberately inclusive. *Invention at Play* challenged "traditional representations of inventors as extraordinary geniuses who are 'not like us.'" The Center instead "celebrate[d] the creative skills and processes that are familiar and accessible to all people."[55]

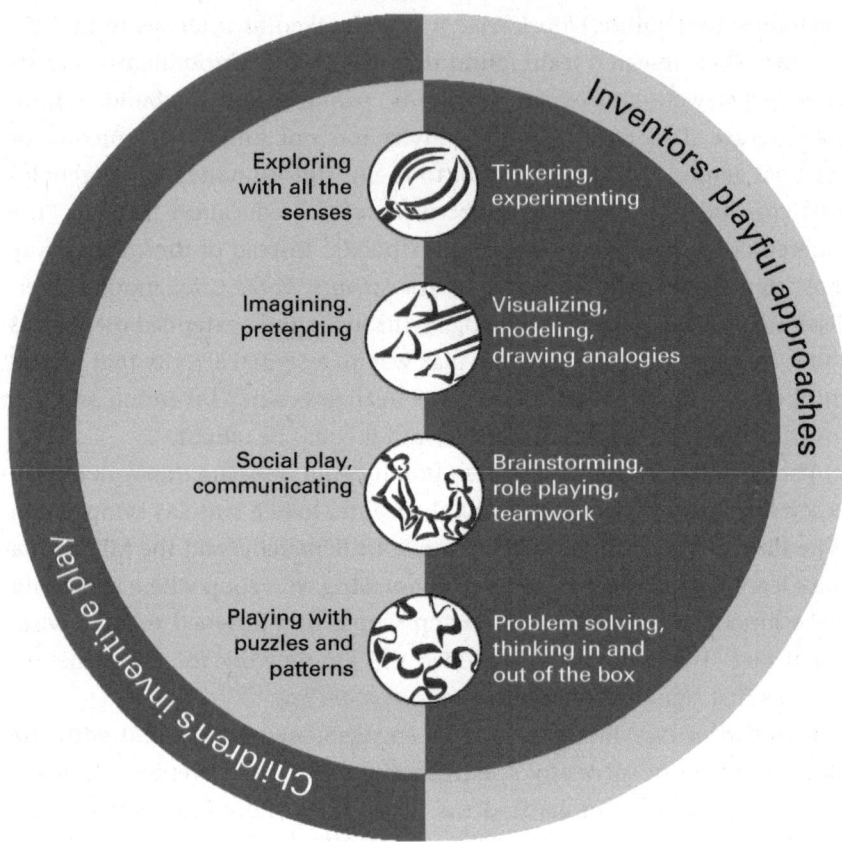

FIGURE 8.4

Links between children's play and inventor's methods. Courtesy of the Lemelson Center for the Study of Invention and Innovation.

The Lemelson Center used *Invention at Play* to integrate and amplify childhood innovation initiatives from across the country. The Center won a $1.6 million grant from the NSF to create the exhibit, which bound together a large network of stakeholders.[56] Collaborators included the Science Museum of Minnesota to help with exhibition development and the Association of Science and Technology Centers (ASTC), a national consortium of museums, to coordinate its national tour. To engage underprivileged kids, it partnered with the YMCA of Metropolitan Washington. Classroom teachers provided feedback on widely circulated curricular materials. Despite its multiple partners, there was no direct corporate sponsorship. The MIT Media Lab was the Center's closest partner. Its Lifelong Kindergarten Group was coming into its own as a leader in technology designed for and with children. With another large NSF grant, the Lifelong Kindergarten Group established the Playful Invention and Exploration (PIE) Network and designed *Invention at Play*'s mobile stations with programmable LEGO "crickets" and musical Play-Doh.[57]

Invention at Play made the case for a bridge between childhood and adult innovation. The first exhibit area, *Playful Inventors*, showcased five exemplary inventors in a series of lessons about how innovation worked. Alexander Graham Bell linked the present to the past and highlighted how the inventor–entrepreneur's telephone resulted from the former teacher's ability to "Borrow from Nature" in his design. In "Keep Making it Better," independent inventor Newman Darby's sailboard demonstrated how sustained curiosity over many years had led to success. Corporate scientist and intrapreneur Stephanie Kwolek's discovery of the bulletproof fabric Kevlar came during a failed experiment, which taught the lesson to "Recognize the Unusual." The inclusion of design firm IDEO highlighted that "Many Heads are Better than One" and that collaboration and multidisciplinary design led to innovative solutions. Finally, McLurkin's robots exemplified how innovative thinkers, like children, "Jump the Tracks" because "they don't know the rules . . . [and] make interesting connections and leaps between different categories of knowledge." These prominent profiles were augmented with inventors and innovators from all walks of life, including Ann Moore, the Peace Corps nurse who designed the Snugli baby carrier, and Krysta Morlan, a high school student who created a physical therapy water bike to help treat her cerebral palsy.

The second area, *The Invention Playhouse*, used hands-on activities to engage visitors in a play toolkit. The Invention Playhouse cultivated "imaginative play" by encouraging visitors to sketch and rearrange magnets in unfamiliar ways. To foster "social play," the Lemelson Center emphasized brainstorming and working together in building and testing models that could be placed in the exhibit's wind tunnel. Finally, kids learned to ask questions and seek new ways of approaching problems using tessellation puzzles.

The third area, *Issues in Play—Past, Present, Future*, contextualized play in a digital world. It used museum artifacts to ask what toys inventors had played with in the past and investigated whether play was changing. This section also included short videos with Kelley, Root-Bernstein, Gopnik, and other innovation experts.[58] All of these activities were linked to national educational standards in the sciences and social sciences.

Invention at Play was enormously successful and inspired several additional initiatives. The exhibit operated for six months at the Smithsonian before traveling to eighteen different science and technology museums across the country over the next decade, and then returned for a second installation at the Smithsonian. The Lemelson Center also developed curricular materials for teachers, students, and the public to cultivate inventive "habits of mind." Finally, *Invention at Play* provided the blueprint for Spark!Lab, the Smithsonian's permanent invention education center which reinterpreted the original aims of the Hands On Science Center with a focus on invention and innovation.[59] The exhibit additionally contributed to the growth and evolution of its many partners. The Lifelong Kindergarten Group, for example, used crickets as a testing ground for the programming language that became Scratch, which today is used by tens of millions of children worldwide. Finally, *Invention of Play* inspired a flourishing of new scholarship that empirically demonstrated the effectiveness of play in delivering informal science education and innovation training to children.[60]

PROFITING FROM PLAY

From Project XL to *Invention at Play*, childhood innovation initiatives created a robust infrastructure to help the nation's youth become innovators. Childhood innovation initiatives sought immediate dividends, measured by kids reached, skills delivered, and attitudes taught. But the ultimate aim of these programs was to foster lifelong growth as their participants scaled Lemelson's pyramid.

In tandem with youth invention programs, innovation experts in the early 2000s argued that *adults* should continue to play like children to maintain their innovative edge. Adult play came with the promise of personal and financial growth. The call was not novel. The "change manager" identity created by the Innovation Group in the late 1960s, for example, made playfulness a virtue in contrast to the analytical nature of Cold War technologists. Celebrations of Silicon Valley millionaires in the 1980s likewise strengthened the association. Calls to change lumbering corporations also seized on the valorization of play as a marker between bureaucratic and innovative organizations. Play was finally crucial to Richard Florida's theory of a creative class, from tech recruiters seeking talent during bouts of ultimate Frisbee to CEO-led rock bands at SXSW.

At the start of the new millennium, innovation experts integrated the lessons of childhood play into toolkits for business innovation. The Root-Bernsteins identified play as one of thirteen vital "thinking tools" used by "the world's most creative people."[61] Daniel Pink, a former speechwriter for vice-president Al Gore, similarly identified play as one of six "essential aptitudes" for success in an innovation age in his best-selling 2005 book *A Whole New Mind*.[62] The *Wall Street Journal* meanwhile described IDEO's company headquarters as "an airy playpen for adults."[63] IDEO leaned into this identity in Tom Kelley's best-selling *Art of Innovation*, which described how the company's "boyish pranks and wild play . . . created an atmosphere where you naturally took chances and solved problems" and recounted how formal events including a soapbox derby and toy workshop fostered "the merging of fun and work."[64] At the MIT Media Lab, business guru Michael Schrage codified these approaches in his book *Serious Play: How the World's Best Companies Simulate to Innovate*. Serious play, Schrage wrote, "is about improvising with the unanticipated in ways that create new value," its aim was not "rediscovering lost childhood," but rather "competitive success" and "innovating profitability."[65]

These business bestsellers built on firsthand experience and academic research in game studies, educational technology, and strategic management. Corporate programming adopted play techniques to shape their hiring, retention, and continuing education.[66] LEGO, for example, turned an internal executive training program into a consulting business via its Serious Play® methodology, which it offered through special LEGO building kits and facilitated meetings.

Outside the workplace, extracurricular play became a potent symbol of the high-tech economy and its creative innovators. In the early 2000s, Burning Man, an experimental art-life festival held in the Nevada desert, was the go-to media example of all that was right and wrong with innovation culture. Initially started as a situationalist art performance, Burning Man grew into a must-attend destination for Silicon Valley elites in which artists and technologists shed their identities and worked together on whimsical and often gigantic creations, culminating with the destruction of a towering effigy.[67] Burning Man in other words acted as a kind of PIE network for adults that reinforced that innovation was a source of playful rebirth.

A FUN DEFICIT

If advocates in the early 2000s were to be believed, play was necessary for unlocking innovation among both children and adults. But were Americans getting enough fun? Was it of the right kind? What would happen if the nation failed to develop its creative capacity?

In the early 2000s, experts merged the rhetoric of an innovation deficit with the imperative for creative play. The premise of business bestsellers like Pink's *Whole New Mind*, for instance, was that because computation and manufacturing could be performed anywhere in the world, Americans would get "left behind" if they did not focus on "aesthetic, emotional, and spiritual demands."[68] Barriers to playful innovation included standardization and rote memorization in schools as well as mindless consumer entertainment on television.

The British education policy expert Ken Robinson emerged as one of the most influential voices for systemic reform to foster childhood creativity for the innovation economy. Throughout the 1990s, he had worked to define the arts as a form of innovation education in the United Kingdom. In 2006, Robinson was selected as one of six thought leaders, along with former Vice President Al Gore, to give the first internet-distributed TED Talks—a playful but serious gathering of thought-leaders. TED's digital strategy succeeded beyond all expectations, and Robinson's lecture "Do Schools Kill Creativity?" was TED's biggest hit. With over 75 million views, it remains the most-watched TED Talk of all time. Robinson distilled the idea of a creativity deficit into a playful, ten-minute masterclass. He argued that the "whole world [was] engulfed in a revolution" in which creativity was the leading currency, but that the forms of education that fostered creativity held a subordinate position. To drive home the point, he told of the challenges and triumphs of playful innovators. For example, it was only through a childhood intervention by a psychologist that *Cats* choreographer Gillian Lynne found her calling. A once fidgety student with difficulty concentrating, "she's given pleasure to millions; and she's a multi-millionaire," he intoned, "Somebody else might have put her on medication and told her to calm down."[69]

A growing maker movement in the United States championed cross-generational play as a form of democratic innovation expertise. The maker movement built on the invention conventions of the 1990s, but eschewed competition for an inclusive and irreverent do-it-yourself ethic. The movement's catalyst was Dale Dougherty, a former technical writer and internet tastemaker of O'Reilly Media. In 2005, Dougherty launched *Make* magazine with the goal of turning consumers into makers. To do so, he profiled "expert makers" and provided the space and blueprints for readers to themselves become experts. Play was the mechanism for achieving this goal. *Make*'s first issue profiled backyard monorails, a LEGO robot for solving Rubik's cubes, and a mechanical pong game.[70] *Make* called for replacing the "mad scientist" stereotype with the image of the "playful scientist" who enjoyed windsurfing for fun.[71] In 2006, *Make* hosted its first Maker Faire for Silicon Valley hobbyists. The events spread rapidly across the United States and around the world with an emphasis on family friendly fun.

The maker movement tapped into desires for play as an alternative to corporate innovation and as one more corporate tool in the battle for global competitiveness. This contradictory dynamic fueled the movement's rise. What started as a hobbyist outlet for adults evolved into significant educational reforms. Elementary schools incorporated arts into the math and science curriculum via a STEAM model to make STEM education more creative, while middle and high schools reinvented shop classes to emphasize craftwork and playful inventing. *Make* magazine was an important broker between stakeholders with differing ideals and ends. Its inner circle included anti-authoritarian writers and internet personalities such as Mark Frauenfelder, Cory Doctorow, Xeni Jardin, and Bruce Sterling. At the same time, technology corporations from Amazon to Disney were the largest sponsors of its Maker Faires. This convergence and legitimation culminated in 2014 when President Obama hosted a Maker Faire on the White House lawn. At the same time, a coalition of mayors across the United States promoted the Mayor Maker Challenge that tied making to a manufacturing "comeback" in which "innovation is helping drive our economy." This democratization of innovation, the Challenge contended, began by bringing children and families together with entrepreneurs, "libraries, museums, universities, schools, philanthropists and community-based organizations" as well as "labor unions" in a movement that was "empowering Americans—young and old—to become the producers of things, not just consumers of things."[72]

The imperative for children to become future innovators made parents important agents in the spread of innovation expertise. To ensure that their children succeeded in life, parents were encouraged to purchase the right toys and enroll their kids in extracurricular STEAM activities. *Parenting* magazine, for example, gave readers tips on how to "Raise the Next Steve Jobs." It simplified the innovator's toolkit to a series of aphorisms including "think different" and "love what you do" while using a child model dressed as Jobs to sell New Balance shoes and Levi's jeans.[73]

These childhood initiatives were accompanied by broadened images of innovators. While *Parenting* used the Jobs stereotype to lure readers, the magazine presented an inclusive pantheon from Albert Einstein to Oprah Winfrey, Jay Z, Gloria Steinem, and Toni Morrison, such that every parent could help their child find an innovative hero. The Innovator Series produced by Kidhaven Press similarly profiled "pioneers" that included Apple designer Jonathan Ive, poet Shel Silverstein, director Tim Burton, Barbie doll inventor Ruth Handler, and Nintendo game designer Shigeru Miyamoto.[74]

Although less didactic, Hollywood was the most influential distributor of aspirational images of innovators aimed at children. The hero of Marvel Studio's 2008 *Iron Man* cast the superhero as an irreverent tech genius whose self-acquired powers and newfound responsibilities didn't dampen his playfulness. The purest expression of childhood innovation came in Disney's animated hit *Big Hero 6*, released at the height of the maker movement, of which Disney was itself a major sponsor. *Big Hero 6* mixed superhero fantasy with the exact qualities that the Smithsonian's Lemelson Center had found in James McLurkin. Its protagonist Hiro Hamada is a child prodigy who invents a microrobotic swarm. At the San Fransokyo Institute of Technology, Hiro meets a likeminded group of playful and diverse innovators. The group bands together to battle corporate greed, the violence that stems from competitiveness, and the misuse of power to instead embrace innovation as helpful, fun, communal, and care based. Innovation, it warned, too often has a dark side; but, if innovators embrace their playful spirit, they could save the world.

CRADLE TO GRAVE

Efforts to turn the youngest of Americans into innovators expanded innovation culture to new levels. Childhood innovation initiatives enrolled classroom teachers, government bureaucrats, equity reformers, multinational

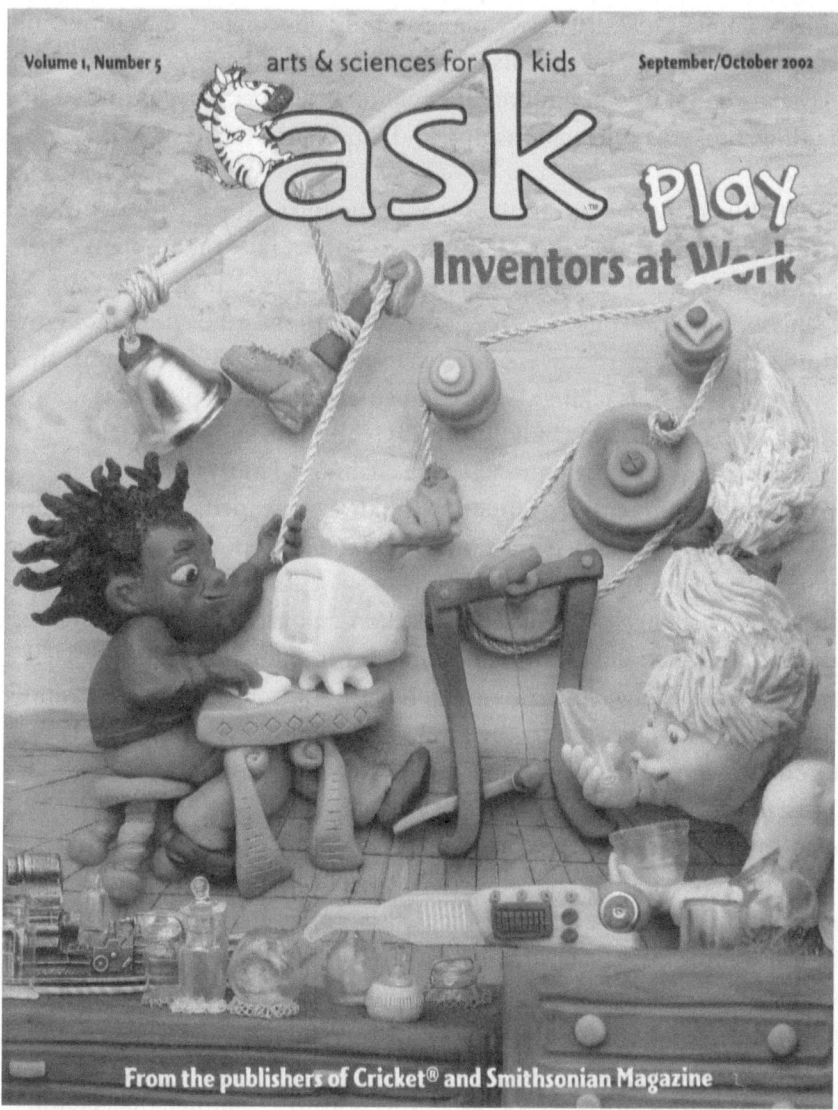

FIGURE 8.5

Inventors at ~~Work~~ Play. The Lemelson Center's *Invention at Play* recast innovation as a multicultural activity in which everyone could participate that linked childhood play to joyful adult work personified in this children's magazine promoting its efforts.

firms, computer scientists, psychologists, museum curators, magazine publishers, toymakers, and parents. Philanthropic foundations, new and old, played a central role in binding this coalition together. They provided the funding, brokered the relationships, and legitimated the ideas as they took shape.

Childhood innovation initiatives were motivated by a two-part thesis; first that there was a need to cultivate exceptional talent that might otherwise be stifled, and second that *all* children whatever their race, class, or economic position could become contributors to the innovation economy. Thus, child's play was always connected to adult work. It was no surprise that *Invention at Play* was taking shape, companies were hiring play consultants, traveling to festivals, and building offices with slides, ping-pong tables, and toy workshops to emulate the "creative chaos of a kindergartener's life."[75]

As innovation entered childhood, however, experts faced tensions between corporate and democratic values. It was neither palpable nor appropriate to foster cut-throat six-year-old entrepreneurs. Experts also faced the challenge of an adult innovation culture that, despite ongoing reforms, was often hostile to women and underrepresented minorities. Play thus provided an age-appropriate and inclusive introduction to innovation culture. Play emphasized creative exploration for its own sake, sometimes through competitive games, but more often through "messing around." The Lemelson Center's *Invention at Play* exhibition distilled and distributed the essence of this vision to greatest effect. It presented innovation as a multigenerational endeavor, pursued collaboratively, accessible to all races and genders. The inventors, scientists, engineers, and artists involved in innovation were motivated by play, not work. Moreover, play, and by extension innovation, was universal and inclusive—everybody plays, so anyone can be an innovator.

The most important feature of play as a tool of innovation is that it presented a progressive vision of innovation in a manner that children themselves could practice. Kids didn't just learn about innovation through *Invention at Play*, Scratch, and Maker Faires. Through hands-on activities, they took their first steps on a lifelong journey of innovation.

III SELF-RECKONING

9   INNOVATORS ON TRIAL

Projecting the world-as-it-is into the future several decades, can you see innovation becoming a crime?

—G. William Jones, *Innovation and Other Modern Parables*, 1962

Consider the following profiles of innovators and the means for making them.

First, meet Ayana Johnson, a marine biologist at the Scripps Institution of Oceanography working to combat climate change. The young Black changemaker documented species loss in Caribbean reefs and drew on insights from biology and economics to design a trap that could prevent millions of endangered species from being caught each year. Her innovation promised to restore reef ecosystems and to maintain the livelihood of local fishermen. Featured in the National Science Foundation's 2011 strategic plan *Empowering the Nation Through Discovery and Innovation*, Johnson personified President Barack Obama's policies to harness innovation for American progress. She shared the spotlight with inspiring innovators of all generations, races, and genders whose contributions ranged from smart clothing to the treatment of brain-injured infants. By highlighting Johnson's work, the NSF portrayed the innovation process as a "dynamic" and "global" force for addressing the twin problems of climate change and inequality amid "unpredictable futures in uncertain economic times."[1]

Next, meet Tyson James, a New Jersey entrepreneur and politician also tackling the climate crisis. At just two-and-a-half years old, James was a recipient of *The New Yorker*'s "3 under 3" award. The magazine anticipated that James's innovations would become "very profitable, very fast." He already had federal support. "If this boy is serious," his senator explained, "it could help trillions of people." The young white boy declared that "for my birfday, I wish there was no global warming?" When asked the secret to his success, James mumbled that "I threw the cheese in the sofa."[2] This satire of diapered disrupters skewered a popular genre of innovator profiles such as *Forbes*'s "30 under 30" and *Popular Science*'s "Brilliant Ten." Its showcase of James extended familiar tropes about innovators to absurdity and implied that innovation expertise was empty bluster.

Johnson's and James's competing profiles were a harbinger of a sharp reversal in American attitudes about innovation. Johnson's achievements in *Empowering the Nation* represented the apogee of more than half a century of investment by academia, government, and industry in pursuit of the belief that innovation was necessary and beneficial. In contrast, *The New Yorker*'s send-up of James revealed unease about an ideology whose broken promises had come to infiltrate everyone's lives, contributing to social upheaval and economic inequality.

This chapter documents and analyzes the splintering of the innovation consensus. It first chronicles the height of innovation culture during the Obama era. It then explores the emergence of anti-innovation movements of academics, journalists, and satirists who revolted against "innovation speak," "coastal elites," and "Big Tech." It investigates how this "Techlash" resulted in a reconsideration of innovation's promises, and surveys innovation experts' attempts to rescue their imperative through reformed mindsets and methods. It also highlights reactionary alternatives to innovation led by Peter Thiel, Elon Musk, and conservative champions for innovation. Finally, the chapter analyzes how a widening partisan divide is remaking ideals about where innovation originates and which innovations are desired.

PEAK INNOVATION

At the start of the new millennium, the watchword "innovation" seemed to capture the life-force of the United States. The tech sector did not take long to recover from the bursting of the dot-com bubble. Rather, the digital technologies of Internet 2.0 were an optimistic counterweight to 9/11, the Iraq and Afghanistan wars, and the global economic crisis. As it had been for the first innovation experts, the pace of technological change was undeniable. By 2000, a vast majority of Americans had access to the internet, which in turn facilitated innovations in mobile computing and social media. Just five years after the Apple iPhone hit the market, two-thirds of Americans owned a smartphone while Facebook grew from a few hundred Ivy League students to over a billion users.[3] Like automobility, electricity, and computing in eras past, the smartphone and social networks came with a magical aura; a tactile merger of technology and culture in the palm of your hand that made all the world your friend. In contrast to the utopianism of the 1950s, these transformations were not attributed to the inevitable march of technology. Rather, journalists, politicians, academics, and PR specialists linked these breakthroughs to an innovation culture that unified technological, economic, social, and political change.

Exuberance for innovation spanned partisan affiliation but was especially prevalent on the left. Before its critical turn, for example, *The New Yorker* was a leading innovation booster. Its inaugural 2007 "Innovators Issue" profiled scientists at CERN's Large Hadron Collider, the graffiti artist Banksy, the architects of New York City's Highline urban park, and billionaire Richard Branson's foray into biofuels.[4] Staff writer Malcolm Gladwell meanwhile became an innovation guru for the masses. His bestseller, *The Tipping Point*, popularized Everett Rogers's diffusion theory, while *Outliers* explored the conditions for innovation and informed parents of the best month to conceive to maximize their child's lifetime success.[5]

Innovators were the heroes of the new millennium. A cottage industry of memoirs and business histories celebrated billionaire founders from Nike's Phil Knight to Amazon's Jeff Bezos and Tesla's Elon Musk.[6] The digital entertainment industry championed artistic creativity, lauding the animator–entrepreneurs of Pixar and hip-hop mogul Dr. Dre who invested in an arts-based innovator academy.[7] The reality show *Shark Tank* played its part by turning pitch meetings into household entertainment, spotlighting entrepreneurs from all walks of American life.

One man, Steve Jobs, served as the national archetype of innovation. Since the early 1980s, Jobs had been a media fixture, at turns a mercurial entrepreneur and failed tyrant. Apple's reinvention from a computer company into a consumer electronics and entertainment firm again transformed Jobs's persona. With his signature black turtleneck and salt-and-pepper beard, Jobs was proof that innovation wasn't just for the young or just about technology and business. When Jobs suffered an untimely death from pancreatic cancer, the tech mogul was enshrined as the greatest innovator to ever live. *The New Yorker* lovingly canonized him at heaven's gate with St. Peter searching the Book of Life on an iPad.[8] Walter Isaacson, president and CEO of the Aspen Institute for Humanistic Studies, published a biography of Jobs just two weeks after his death that placed the Apple CEO in the Pantheon of American heroes, declaring that he was "a magician genius" who "combin[ed] the power of poetry and processors."[9]

Barack Obama's presidency was as crucial to this innovation triumphalism as Silicon Valley's cultural and economic dominance. Obama was the ideal type of the social innovator. His rise to power bore the hallmarks of Rogers's theory: a comparatively young, cosmopolitan outsider with a message of optimism and change. His 2009 Inaugural Address championed "the risk-takers, the doers, the makers of things," as drivers of American history.[10] Technology journalists reinforced this charismatic image, labeling Obama the "entrepreneur-in-chief" and "innovator-in-chief."[11] Innovation was central to Obama's governing philosophy.[12] His Strategy for American Innovation fused the technoeconomic blueprint of the 1980s with a focus on access and inclusion. He filled his cabinet and federal agencies with interdisciplinary experts who moved back and forth between government and the high-tech sector.[13]

With the zeal of Herbert Hollomon in the 1960s, bureaucratic innovators in the Obama administration expanded upon decades of prior initiatives. In a suite of programs reminiscent of Jerome Lemelson's National Plan, the Obama administration cultivated innovators at every life stage. At the top of the pyramid, the White House increased the number of National Medals of Technology and Innovation awarded to women, Black, and immigrant innovators.[14] At the midlevel, the Obama administration created a bevy of new programs. The White House Presidential Innovation Fellows recruited bureaucratic innovators to reform government services with digital transparency. The NSF Innovation Corps (I-Corps) provided a "boot camp" that fostered a "start-up mentality" among academic scientists and engineers. NASA and DARPA used open innovation prizes and crowdsourcing to harness the innovative capacity of scientists and everyday citizens. At the college level, the NSF-funded University Innovation Fellows program trained design thinkers at over two hundred colleges nationwide.[15] Finally, the

FIGURE 9.1

Innovator in Chief. President Barack Obama and Education Secretary Arne Duncan talk with students while visiting a classroom at the Pathways in Technology Early College High School (P-TECH) in Brooklyn, NY, Oct. 25, 2013. Official White House Photo by Pete Souza.

Educate to Innovate initiative extended the push for creative innovators in the nation's K–12 public schools, which it celebrated at White House science fairs and Maker Faires.[16]

Obama's innovation programs championed social progress and empowerment but relied on the same deficit fears that had structured innovation policy since the 1980s. In his 2011 State of the Union address, Obama argued that "encouraging American innovation [was] the first step in winning the future."[17] Think tanks, private foundations, and professional associations likewise pushed for a muscular innovation policy to remain globally competitive. The 2015 manifesto *Innovation: An American Imperative*, for example, warned that "competitor nations" were "copying our playbook for success" and would "soon surpass the United States as the global innovation leader."[18]

Deficit arguments had become shopworn, but they were increasingly accurate. Other nations *had* emulated the innovation playbook.[19] In 2010, the European Union launched its flagship Innovation Union initiative, which like *Empowering the Nation* blended technoeconomic innovation with social and environmental reform.[20] Individual nations similarly identified innovation as vital to their prosperity. President Emmanuel Macron, for instance, declared France to be a "start-up nation" and pursued systemic regulatory reforms. China and India embraced policies to surpass the United States and overcome their negative stereotypes as labor and tech support.[21] In Latin America and Africa, a combination of indigenous advocacy and external organizations adopted innovation policy as the dominant lens for development.[22]

Innovation experts were visible change agents in this global imperative. Rogers's diffusion model had become ground truth for entrepreneurs in the dot-com and social media booms. At Stanford's d.school, David Kelley promoted design thinking as applicable to any problem, in any domain, anywhere in the world. In a tenth anniversary edition of *The Rise of the Creative Class,* Richard Florida charted the ongoing accuracy of his initial predictions; the United States Census meanwhile adopted his terminology.[23]

Innovation expertise had grown into a diverse and widely practiced vocation. Think tanks proffered their own models, including the Brookings Institution that distributed blueprints for "Innovation Districts" to municipalities across the country.[24] Universities, high schools, and even elementary schools developed innovation and entrepreneurship programs that reached thousands of students. A new generation of innovation experts, such as Henry Chesbrough and Eric von Hippel, came to prominence with models that stressed openness and democracy.[25] Harvard Business School professor Clayton Christensen achieved the widest celebrity for his notion of disruptive innovation, arguing that insurgents on the margins of an industry had the greatest likelihood of producing creative destruction.[26] His key empirical case study of digital memory storage was unsexy, but his toolkit proved remarkably seductive. Start-up founders and community activists alike argued that every stale and oppressive element of the status quo stood to be positively disrupted.

So, who threw the cheese in the sofa?

TECHLASH

The growth of innovation culture was the result of a decades-long persuasion process during which its evangelists encountered only limited resistance. In the 1970s, for example, scientists had criticized the NSF's innovation programs, arguing that the government should stay away from commercialization and stick to fundamental research. Engineering professors likewise had dismissed entrepreneurship education as a marginal activity. Technology journalists had developed the trope of technology entrepreneurs as socially awkward nerds and geeks. In the 1980s and 1990s, the occasional skeptical economist challenged the long-term benefits of regional innovation schemes.[27] After the 2000 dot-com crash, business insiders and late-night television hosts ridiculed misbegotten ideas like Pets.com.[28] On the whole, however, innovation culture amassed power and support.

In the early 2010s, innovation culture experienced its first sustained challengers. Most dissent originated on the academic left. In 2011, the same year that the NSF released *Empowering the Nation* and Isaacson published *Steve Jobs*, a set of influential books criticized the "privatization" of science, the "delusion" of technology fixes to societal problems, and the "corrosive" psychological and social impacts of entrepreneurial work.[29] The Oscar-winning movie, *The Social Network*, meanwhile probed the dark side of innovators, dramatizing Facebook CEO Mark Zuckerberg's rise to power.

Still, innovation skepticism remained a fringe position.

During President Obama's second term, challenges to specific elements of innovation culture coalesced into a synthetic critique. Leftist intellectuals again led the attack. In 2014, the *New Republic*'s Evgeny Morozov called innovation a "naive fetish" and pointed to the concept's bipartisan support as an indicator of its vapidness.[30] In 2015, technology historian Lee Vinsel cast himself as a digital-age Martin Luther in his biting essay on "95 Theses on Innovation."[31] At the height of the 2016 presidential campaign, populist critic Thomas Frank excoriated liberalism's "excessive hope" and called Obama a puppet of the "Innovation Class." According to Frank, this "rapacious" cult rationalized capitalist inequality through its worship of "credentialed expertise," meritocracy, and education as the basis of social change.[32] A series of digital magazines including *Baffler*, *Aeon*, and *Jacobin* focused their critiques on innovation culture's flawed ideology and its detrimental impacts on labor.[33] Inside the academy, historian Benoît Godin established the journal *NOvation* for the growing specialty of "critical innovation studies."[34]

This innovation revolt was part of a broader Techlash that revived the antitechnology genre of the 1960s. The Techlash linked the digital order to surveillance, racism, environmental destruction, and the death of democracy.[35] Scholars, investigative journalists, disenchanted technologists, and activists published hundreds of articles and dozens of books critical of innovation culture. Perhaps nowhere was this debate more prominent than on the social media platforms such as Twitter that the critics simultaneously used to gain a following and to target.

The entertainment industry also turned against innovation. The HBO satire *Silicon Valley*, for example, began in 2014 with a cast of nerdy down-on-their-luck innovators out to slay corporate giants; but as the series unfolded, it revealed innovation culture's endemic rot and made the case that every participant was complicit. By the end of the 2010s, there was a thriving commercial market for docudramas that included *The Social Dilemma*, *We Crashed*, and at least four exposés on disgraced Theranos founder Elizabeth Holmes.

As the country reached new levels of partisan rancor during the 2016 election, "Big Tech" became a shared epithet of the left and the right. Critics with opposing politics shared the same insults, comparing innovation culture to drugs and venereal disease.[36] Critics on the left sought to dismantle the monopolistic power of the "tech giants," their tax avoidance, and inhumane working conditions. They also accused social media companies of putting profits over people, resulting in lax security and misinformation that enabled Russian hackers to sway the 2016 election. Rightwing politicians and pundits meanwhile railed against "woke capitalists" in innovation regions that had spread from California and Massachusetts into Texas, Colorado, Georgia, and other conservative states.[37] They attacked the tech industry's use of H1-B visas as a form of sweatshop labor and amnesty for immigration.[38] The conservative media outlet *The Federalist*, for example, published over five hundred articles on how "Big Tech Oligarchs" corrupted children, colluded with China, stifled competition, censored freedom of speech, hurt public health, and supported terrorism.

Donald Trump was the Techlash's disruptor-in-chief. At first, President Trump's innovation policy appeared to continue the pattern of democratic and republican administrations before him. In 2017, for example, Trump announced a new White House Office of American Innovation (OIA) to "bring together the best ideas from Government, the private sector, and other thought leaders to ensure that America is ready to solve today's most intractable problems."[39] However, through a combination of mismanagement and deliberate dismantling, the Trump administration reduced the federal government's innovation infrastructure. Foregoing innovation experts, Trump appointed his son-in-law Jared Kushner to lead an OIA staffed with political cronies.[40] The Office of Science and Technology Policy (OSTP), which had acted as Obama's innovation brain trust, dwindled from a staff of 135 to less than 30.[41] Trump also failed to award a single National Medal of Science or National Medal of Technology and Innovation. Finally, as both Trump and Silicon Valley became toxic to one another, several Silicon Valley leaders such as Apple's Tim Cook shifted from quiet collaboration to avoiding association with the White House altogether.

The development of a COVID vaccine, initiated by Trump's Operation Warp Speed, troubled easy narratives about the party politics of innovation. The rapid development process brought together biotech start-ups, corporate giants, and federal, state, and local governments. Yet its roll-out, adoption, and widespread resistance became the most visible symbol of an innovation divide with life-and-death consequences.

The Techlash also brought a dramatic reversal in representations of innovators. Popular media replaced glowing profiles of entrepreneurial heroes with takedowns of greedy and rapacious villains. There were *lowlifes* such as hedge fund "pharma-bro" Martin Shkreli, whose business model was to purchase out-of-patent drugs and raise their prices by as much as 5,000 percent. There were *disgraced disrupters*, like ride-sharing Uber cofounder and CEO Travis Kalanick who was forced to resign amid revelations of ethical violations and sexual harassment. There were *frauds*, like Theranos CEO Elizabeth Holmes, who *The New Yorker* had once praised for "democratizing healthcare" with her rapid and inexpensive blood tests but who had deceived patients and investors from the beginning. Finally, there were *robber barons* and *supervillains*—Jeff Bezos, Mark Zuckerberg, and Elon Musk—whose achievements were as undeniable as they were unavoidable. These billionaires had amassed so much wealth and power that they existed above the law and normal society.[42]

WHAT'S WRONG WITH INNOVATION?

After World War II, experts in multiple fields had celebrated innovation as the answer to the biggest questions of a postindustrial society: How are new things created? How do ideas spread? Where does wealth come from? How can the world be changed? The Techlash challenged these pillars of the innovation consensus. Where innovation experts saw a harmonious synthesis of technology and culture, critics identified a destructive neoliberal order.

At the core of the innovation imperative was a belief in the society-changing power of digital technology. In the late 1950s, technology managers at Bell Laboratories promoted innovation to describe the blurring of science, technology, and society that had given rise to the transistor and, along with it, hybrid technoscientific careers. Engineers, entrepreneurs, and journalists carried this vision westward, portraying Silicon Valley as the leading edge of a decentralized, creativity-based future.

In contrast, anti-innovation experts in the 2010s argued that the relentless pursuit of digital technology had resulted in a misallocation of the nation's talent. Critics argued that from kindergarten to college, the American education system overemphasized STEM careers at the expense of well-rounded citizens. Addictive social media networks brought record users and profits while the nation's infrastructures crumbled. Designers built gadgets for the innovation class, while residents of Flint, Michigan, drank poisoned water. Meanwhile, memoirs by disillusioned insiders criticized Silicon Valley's cutthroat competition, magical thinking, risky technologies, and propensity for fraud.[43]

A second promise of the innovation consensus was that its experts could accelerate the uptake of desirable social and technological changes. From Rogers to Florida, innovation experts offered models that capitalized on the relationship between ideas and social and material change.

During the Techlash, however, critics accused these gurus of peddling a bankrupt ideology. The expert takedown became its own genre in which "delusion" became the newest buzzword.[44] Urban theorists accused Florida of selling false hopes to struggling regions and normalizing the pathologies of booming cities.[45] Science and technology studies scholars declared design-thinking to be an imperialist ideology.[46] "Disruption" was the dirtiest term of all. In *The New Yorker*, historian Jill Lapore attacked her Harvard colleague Christensen, calling his vision of disruptive innovation a misguided "gospel."[47] The rhetoric of religious conflict was a common framing; innovation culture was a cult, its leaders were corrupt prophets, and its aspiring innovators were brainwashed disciples.[48]

Another foundational claim of the consensus was that innovation was the driver of economic growth. Ever since economist Robert Solow had identified technical change as the secret to productivity, bureaucrats, corporate boosters, and politicians advocated for the right mix of regulatory reforms and government incentives to foster entrepreneurial activity, industrial policy, and regional clusters.

In the 2010s, a range of critics instead assailed innovation policy as an engine of inequality. Rather than a perpetual flourishing of start-ups, they claimed that the digital economy produced unregulated monopolies that throttled competition. The resulting concentration of wealth and power produced a stratified economy with billionaires at the top and an expanding service class at the bottom.[49] Innovation hubs like San Francisco, Boston, Seattle, Austin, and Washington, DC, became playgrounds for the rich that were unaffordable for anyone else. Competition for talent in universities was fierce, resulting in a similar divide between winners and losers. However, despite forty years of efforts to diversify the innovation economy, racial and gender discrimination persisted.[50] These inequities at every level strained the credulity of claims that social and technological progress proceeded in harmony.

Critics further challenged the productivity gains ascribed to innovation. The economist Tyler Cowen made the case that the United States had reached a "great stagnation."[51] In *The Rise and Fall of American Growth*, economist Robert Gordon, a former student of Solow's, added credibility to the argument, showing that the rate of overall innovation had been in decline since the 1970s.[52] The conservative pundit Ross Douthat amplified the message. Asking "Where have you gone Thomas Edison," he concluded that "twenty-first-century growth and innovation are not at all what we were promised that they would be."[53]

Even innovation's champions identified worrisome trends. In 2013, for example, MIT assembled a task force to revisit its Reagan era *Made in America* project. The *Making in America* report documented the ongoing decline of manufacturing despite decades of public–private initiatives and lamented the persistence of regional inequality. "Our own initial lofty conceptions of innovation," the MIT Commission concluded, had been "brought down to American earth."[54]

The final, unifying pillar of the innovation consensus was that its mindsets and methods were a path to civic renewal. Ever since the democratic experiments of John Gardner in the 1960s, experts had claimed that public–private alliances could change lives and build meaningful relationships.

These programs focused on education reform and community development with the goal of extending innovation's benefits to those previously excluded.

Critics of civic innovation, however, found dashed dreams and unfulfilled promises at every level. In 2017, after participating in a "progressive" project to create a public school for underserved students via a collaboration between educators and game designers, the anthropologist Christo Sims offered a scathing picture of failure.[55] He argued that a fetish for technological solutions reinforced the status quo and that the "professional fixers" behind innovation-centric programs used a toxic form of idealism to ignore their failures while continuing to spend taxpayer money. *Stanford Social Innovation Review*'s David V. Johnson similarly argued that the Obama administration's courting of young technologists into government service created a "revolving door for our new Establishment."[56] At the global level, former management consultant Anand Giridharadas attacked philanthropic foundations as tools for burnishing the reputation of innovation culture's corporate winners.[57]

The general conclusion among critics was that innovation was, at best, a Pollyannaish strategy doomed to fail and, at worst, a deliberate strategy for maintaining the status quo.

DON'T BE EVIL

The Techlash sparked an expert reformation inside innovation culture. Some champions at first tried to ignore the critiques and later offered impassioned defenses.[58] Other experts, like geocentric astronomers during the Copernican revolution, incorporated a growing list of anomalies into their models.[59] Finally, a variety of reformers resolved to apply innovation expertise to fix innovation itself. This resulted in the development and expansion of several specialist communities.

SOCIAL INNOVATION AND ENTREPRENEURSHIP

Social innovators and entrepreneurs argued that the same methods that produced gains in the tech sector could be adapted to solving problems of poverty and food insecurity. This social innovation approach predated the Techlash, stemming from initiatives that emerged in the Reagan era.[60] In the 2010s, however, social innovation and social entrepreneurship were among the fastest growing specialties of innovation expertise even as critics attacked them as neoliberal instruments.

The first social innovation organization, Ashoka Innovators for the Public Good, was founded in 1980 by Bill Drayton, a former McKinsey & Company consultant and Environmental Protection Agency administrator. With the support of a MacArthur Foundation genius grant, he turned a part-time passion project into a global network. Named after an ancient Buddhist ruler who abandoned militarism for peaceful growth, Ashoka had a two-part mission: make the social innovator a recognized career path and provide seed funds for these change agents to realize their ideas. By 1992, Ashoka had financed over three hundred individuals, such as Brazilian graduate student Mary Allegretti who organized Indigenous communities to fight rainforest deforestation and Mexican housing-rights organizer Antonio Paz who founded Campamentos Unidos (tent-dwellers united).[61]

Another former McKinsey consultant, Gregory Dees, helped make social entrepreneurship education into a global movement. In the 1980s, Dees left the consulting rat race to earn a PhD in applied ethics. He then launched the Social Enterprise initiative at the Harvard Business School to convince students destined for the Fortune 500 to instead pursue careers in social change. Dees combined instruction on venture formation with techniques for building community self-reliance, reducing poverty, and fighting for environmental justice.[62] After a sabbatical working in rural Appalachia, Dees moved to Stanford where, in 2003, he cofounded the Center for Social Innovation and helped launch the *Stanford Social Innovation Review*.

Philanthropic foundations in the early twenty-first century followed the lead of these first movers, embracing social innovation as a method and a mission. The Aspen Institute created the first of its many social innovation projects in 1993 with the rationale that it was a moral good but also a practical necessity.[63] Silicon Valley's Social Venture Network similarly advocated for repairing cultural, economic, and racial divides in the Bay Area through investments of "virtuous capital."[64]

In the 2010s, business schools around the world created dozens of social innovation and entrepreneurship programs. Magazines augmented their "top innovator" lists with social entrepreneurs.[65] The Lemelson Foundation also branched into social innovation, partnering with the AAAS to support "global invention ambassadors" who targeted disease and environmental

FIGURE 9.2
"Everyone a Changemaker." Courtesy of Ashoka.

sustainability while "serving as living examples of what an inventor can accomplish."[66]

Social innovation advocates redefined the innovator as a changemaker whose primary tools were empathy and community-building. On Ashoka's twenty-fifth anniversary, the organization introduced the new tagline: "Everyone a Changemaker," and later launched "changemaker campuses" at universities, high schools, elementary schools, and museums across the country. The *changemaker* bore many of the traits of the *change manager*. Both put innovation expertise front and center. Both emphasized the power of ideas. Both blended nature and technology, youth and age. And both had faith that collaboration could change the world. The changemakers, however, were a nonprofit, multiracial team who insisted that everyone could contribute; in fact, good innovation only happened when everybody did.

INCLUSIVE INNOVATION

Where social innovators applied their toolkit to heal global divides, a second cluster of reformers confronted inequities *inside* innovation culture. These inclusive innovation advocates built on decades of past diversity initiatives as they worked to make innovation culture both equitable and profitable.

Reformist economists defended innovation as a driver of economic and social progress, but called out the underrepresentation of women, minorities, and low-income people among innovators. An influential 2017 study, for example, found that children raised in high-income households were ten times more likely to be inventors and innovators compared to the median. Similarly, children who grew up in innovative regions like Silicon Valley were much more likely to become innovators than kids in underresourced communities. This lack of exposure along lines of race, class, and gender resulted in "lost Einsteins."[67]

Echoing Rosabeth Moss Kanter's work from the 1980s, consultants claimed that diverse teams led to more innovative outcomes. Meanwhile student initiatives such as Black Girls Code attracted corporate and philanthropic sponsors and enthusiastic students. Other reformers addressed issues of identity and power. Olin College engineering professor Debbie Chachra, for example, criticized the maker movement's masculine culture and advocated for replacing it with an ethics of care.[68] The group 500 Women Scientists, which formed in response to Trump's 2016 election, argued that science "fuels innovation" only if it is harnessed to fight "racism, patriarchy, and oppressive societal norms."[69] Stanford University's ambitious Gendered Innovations project worked with federal agencies in the United States and the European Union to demonstrate the benefits of feminist and antiracist perspectives in enhancing innovation and to transform how scientists, engineers, entrepreneurs, and funding agencies conducted research and developed new technologies. It cultivated a feminist toolkit drawn from cases ranging from the use of pregnant crash-test dummies for safer cars, to the analysis of sexist chatbots as an example of how not to innovate.[70]

RESPONSIBLE INNOVATION

A third group of reformers aimed to alter innovation culture by changing its methods. This *responsible innovation* movement coalesced in Europe and the United States among policy experts.[71] Its advocates argued that new forms of sociotechnical expertise could better regulate innovation and mitigate its harmful impacts.

Responsible innovation experts began with the recognition that innovation was a mixed blessing with differential benefits, risks, and outcomes.[72] In a democratic society, they argued, everyone should have a role in shaping innovation so that its fruits were more equitably distributed and its harms reduced. However, the complexities of a technoscientific society demanded experts who understood how that society functioned. Responsible innovation advocates viewed themselves as honest brokers who could help achieve change by mediating among stakeholders and fostering reflection about the innovation process. Their methods included *anticipatory governance* to evaluate the future impacts of an innovation before deciding to pursue its development, *midstream modulation* that utilized evaluation and self-criticism evaluation during the R&D process, and *stakeholder dialogue* so that innovation policy included public input.[73]

In the United States, responsible innovation's leading change agent was the political scientist David Guston. Guston's career bridged the technology assessment movement of the 1970s with the innovation policy of the post–Cold War era.[74] Guston had been an intern at the Office of Technology Assessment before Congress eliminated the agency in 1992 and had studied as a postdoctoral fellow under the direction of the technology–economics expert Lewis Branscomb. Guston made his name during the government's National Nanotechnology Initiative (NNI), a $36-billion-dollar effort to catalyze innovation and global competitiveness in a frontier technoscientific field.[75] In the early 2000s, he helped secure millions of those billions to study the societal dimensions of emerging technology, which came with utopian promises but stirred dystopian fears about runaway swarms of nanobots.[76] He and colleagues established the Center for Nanotechnology in Society (CNS) at Arizona State University, an institution that was making headlines for its promises to democratize innovation.[77] Although the nano-future never arrived, investments in CNS resulted in a generation of new tools and experts. In 2014, this hybrid community established the *Journal of Responsible Innovation*. The following year, CNS became the School for the Future of Innovation in Society with the tagline "The Future is For Everyone."

A wide network of responsible innovation programs aimed to train new generations of ethical innovators. Fellowships targeted existing practitioners in industry and government. The Washington, DC, philanthropy New America, for example, established the Public Interest Technology network with over fifty university members to channel research and education toward responsible design. In Silicon Valley, venture capitalist Roger McNamee created the Center for Humane Technology, which offered a self-paced course for innovation professionals to "build technology that matters tomorrow" utilizing "different principles today."[78] In 2016, MIT engineer Christine Ortiz collaborated with technology historian Ellan Spero to create *Station1* to train underrepresented college students to be masters of "socially-directed science and technology."[79]

DESIGN JUSTICE

A related *design justice* movement embraced the power of innovation as a means of social change but was explicit about the exclusionary failures of a system that they argued was rooted in colonialism. The movement took its mantra from disability rights activists: "Nothing about us without us." Coming from BIPOC, LGBTQ, and marginalized communities, its advocates sought to build a world in which everyone could achieve their "full self" through creative labor. Its proponents shared an unease with "genius" and "savior" tropes and instead emphasized collaborative, less ego-centric personas of "practitioners" and "coalitions."[80]

Advocates argued that design justice was a kind of "anti-expert" expertise that through radical inclusion sought to decenter who could and should innovate.

The design justice movement emerged in the context of community organizing, human-centered design, and the new-media cultures that Florida had identified as havens for "super-creatives."[81] As a conceptual framework, design justice linked a series of overlapping efforts, including feminist and anti-capitalist pockets of the maker movement; critics working to "decolonize" design; disability advocates fighting against techno-ableism; designers who integrated political theory into *adversarial, equity-centered,* and *liberatory* methods; and professionals who demanded control over their labor, such as the Tech Workers Coalition.[82]

The movement's main incubator was the Allied Media Conference (AMC), an annual gathering of activists in the arts, media, communications, and technology. Since AMC's founding at the turn of the millennium, the Detroit-based event had come to attract over two thousand participants annually. In 2014, a group of attendees created the Design Futures Group to imagine alternatives and to "begin making them into reality."[83] In 2016, the group announced the Design Justice Network as a way of moving beyond "design for social impact" and "design for good," which, it argued, perpetuated the problems that practitioners aimed to solve because those practitioners did not question their privilege. Design justice advocates wanted to give the power back.

One of the network's key nodes was ironically one of innovation culture's leading centers of power. In 2016, MIT Media Lab director Joi Ito announced a $250,000 Disobedience Award for exceptional social and political disruptors to "harness responsible, ethical disobedience aimed at challenging the norms, rules, or laws that sustain society's injustices."[84] Its first recipients were medical doctor Mona Hanna-Attish and engineer Marc Edwards, who exposed the Flint, Michigan, water crisis.

Media Lab professor and trans rights activist Sasha Costanza-Chock used their institutional position to amplify the design justice movement's message.[85] In 2020, they published *Design Justice: Community-Led Practices to Build the Worlds We Need*, a book that established Costanza-Chock as the face of a decentralized movement and made them a sought-after innovation consultant who offered workshops so that "key people," including inside Big Tech companies, could learn Design Justice Network Principles, with the proceeds funding the network's growth.[86]

The Design Justice Network's main toolkit was its set of ten Principles. The first identified design as a generative means to "*sustain, heal*, and *empower* our communities" as well as a critical weapon "to seek liberation from exploitative and oppressive systems." Other principles resonated with the T-groups of the 1960s and the aims of the Innovation Group; for example, that change was an emergent and collaborative process "rather than as a point at the end of a process." However, these design justice principles were wed with an emphasis on accountability and accessibility and that the designer was a "facilitator" rather than an "expert."[87]

MAINTENANCE, CARE, AND REPAIR

Finally, a group of critics worked to realign an innovation-obsessed society toward alternative values of maintenance, care, and repair. This *maintainers movement* began in opposition to innovation culture but was both developed and embraced inside its institutions.

As with other innovation reform movements, advocacy for maintenance, care, and repair predated the Techlash. The maintainers movement originated around several themes that included an ethics of care, a philosophy rooted in feminist psychology that prioritized empathy and compassion; "broken-world thinking," which identified repair as a societal imperative but rejected technological fixes; theories of a circular economy aimed at reducing capitalist growth; and histories of technology that demystified the role of innovation in social change.[88] The movement additionally linked DIY Repair Cafés and Right-to-Repair consumer advocates who challenged the proprietary ownership and forced obsolescence models of technology companies.

In 2016, a viral essay by technology historians Andrew Russell and Lee Vinsel, gave the maintainers movement its name and identity. In "Hail the Maintainers!" they argued that the worship of innovation had created a false picture of the technological world that led to the neglect of essential maintenance and repair and reduced the people who perform that vital work to a subordinate status.[89] They drew attention to the older infrastructures and legacy technologies of electrical grids, highways, and sewer systems obscured by the obsession with the new. Moreover, they argued that innovation culture's cherished value of disruption worked in opposition to the stability and reliability of critical infrastructures. Russell and Vinsel rejected the persona of the innovator, instead championing the invisible and underappreciated workers who kept physical and digital infrastructures running—urban waterworks administrators, IT professionals, nurses, auto technicians, and Air Force mechanics.

The argument was a potent rallying call for anti-innovation critics and drew bipartisan attention to issues of maintenance, care, and repair. With funding from the innovation-centric philanthropies the Alfred P. Sloan Foundation and the Siegel Family Endowment, Russell and Vinsel worked with the nonprofit Educopia to create the Maintainers Network. The network hosted a series of research workshops and established a fellowship program to develop a "community of practice" that supported scholars and practitioners in design, urban planning, public health, art, and ecology.

DESIRABLE, BUT VIABLE?

Across the Trump years and the global pandemic, reformers concluded that innovation was no panacea. To change the culture, they sought to remake power relations. To remake power relations, they appropriated, reformed, and applied mindsets and toolkits of innovation expertise. In a project made familiar by seventy-five years of effort, they once more sought to define the identities of who could innovate. Reformers also increasingly delimited who *should not*.

Reformers' attempts to disrupt the dominant culture, however, encountered a series of challenges faced by any new venture.

Sounds good, but will it yield results? As reformers critiqued innovation culture's dominant ideology, each movement wrestled with its own ideology and the difficulties of measuring success. Responsible innovation emphasized complex theoretical tools and acronyms for policy insiders. Design justice deployed a vocabulary with high start-up costs for those not in the social justice community, as well as concern about the balance between the "productive" and "disempowering" role of critique.[90]

Is the idea as innovative as promised? Reformist movements all advocated for some variant of democratizing innovation. Nearly all, however, drew financial, institutional, and reputational support from leading innovation organizations. The social-, inclusive-, and responsible-innovation movements were intentional about changing the system from within and therefore sought the most powerful partners possible. While reformers fashioned their initiatives in "productive conflict" with the establishment, critics dismissed them as defanging true alternatives.[91] The design justice and the maintainer movements often emphasized working outside or against innovation culture, but their leaders had also built their careers within it.

Does the team agree on a direction? Despite significant overlap in goals and methods, the array of alternative networks represented multiple visions for changing the status quo. The result was reminiscent of competing entrepreneurship programs in the 1970s with proponents expended energy quarreling with each other. Design justice advocates, for example, positioned their work in contradistinction to social innovation approaches on one flank and defended themselves against the critique of the maintainers on the other.

Can the essence of the innovation be maintained on the path to adoption? Finally, each of these movements struggled with the relationship between their ideas and their tools. Vinsel and Russell's maintainers' manual, *The Innovation Delusion*, for example, was published by a leading pro-innovation firm whose catalog included Thiel's *Zero to One*. The book utilized a familiar rhythm to that of the consultants they critiqued: it celebrated maintenance *experts*, promoted a maintenance *mindset*, spotlighted maintainer *personas*, and invited readers to join a maintainer *community*. The maintenance mindset also had a recognizable melody; it "sustain[ed] success," "depend[ed] on culture and management," "require[d] constant care," and demanded "a long view."[92]

Even the most skeptical investor, however, could not deny the impact of the reformers on innovation culture. Corporations appropriated *frugal innovation* methods developed by social entrepreneurs, while a new class of B-corporations (B for benefit) aimed to make business a "Force for Good."[93] Pro-innovation outlets like TED now feature critics in equal measure to boosters. Tech giants from Alphabet to Microsoft hired—and fired—high-profile ethicists and held workshops with design justice advocates.[94] Philanthropies such as the Mozilla Foundation and the Social Science Research Council created fellowship programs for responsible computer science and Just Tech. Politicians with an eye on 2024 elections, such as California congressman Ro Khanna, blended innovation reform into campaigns that promoted "progressive capitalism" and innovation that "work[s] for us all."[95]

The Biden administration incorporated these demands for repair and justice into federal innovation policy.[96] Its signature achievement, the Creating Helpful Incentives to Produce Semiconductors (CHIPS) Act, fused the competitiveness dreams of the technology–economics group of the 1980s with an emphasis on manufacturing jobs. Pro-innovation think tanks lauded CHIPs—which in early drafts was called the "Bipartisan Innovation Act"—for its potential to "heal the nation's economic divides."[97] The new NSF "Engines" program supported by the act updated the all-encompassing promises of the innovation imperative by "catalyz[ing] robust partnerships to positively impact the economy within a geographic region, address societal challenges, advance national competitiveness and create local, high-wage jobs across the country."[98] Meanwhile the Infrastructure Investment and Jobs Act blended maintenance and innovation with billions of dollars to repair existing infrastructure and to develop new renewable energy technologies with "proven innovative practices."[99] Likewise, in December 2022, Vice President Kamala Harris announced the Equity and Excellence initiative for "empowering and training the next generation of innovators and removing the barriers to these talented individuals achieving their full potential."[100]

RETURN OF THE UNREASONABLE MEN

Liberal and progressive reformers were not the only group seeking to reshape innovation culture from within, nor the only ones to demonstrate results. A different coalition placed the blame for America's innovation failures on the liberal orthodoxy of monopolistic companies. Blurring libertarianism and populism, this reactionary strand of innovation culture defended innovators as outliers who made exceptional contributions to human progress. Like their progressive counterparts, they worked to remake government-industry relationships and to train the next generation of innovators in their image.

During the 2016 election, journalists discovered this "secret republicanism" among tech workers who were "hiding their true selves" to conform to Silicon Valley's liberalism.[101] In the series *Divided We Code*, for example, CNN used witness protection silhouettes to dramatize "undercover conservatives" in the tech sector.[102] After Trump's victory, the conservatives no longer had a need to hide.

Innovation culture's rightward turn was neither a mass uprising nor a simple shift in the zeitgeist. It had historical roots in California's defense industries as well as the white supremacism of microelectronics pioneer William Shockley.[103] In the 2000s, its key agents came from a network of engineers and financiers from the "PayPal mafia" and its founders Thiel and Musk. The peer-to-peer electronic banking company resulted from the merger of their competing start-ups in 2000, and just two years later was acquired by eBay for $1.5 billion in stock. Thiel leveraged PayPal's success to become a top venture capitalist—bankrolling Facebook, Airbnb, and Uber in their infancy—and then a conservative powerbroker. Meanwhile, Musk founded Tesla and SpaceX on his path to becoming the world's wealthiest human. Over the next twenty years, the two founders contributed to a vision of innovation that celebrated transgression as innovation's highest virtue.

According to Thiel's biographer Max Chafkin, the venture capitalist's "contrarian" philosophy derived from life as a Silicon Valley insider who always felt on the outside. From a young age, Thiel had been an awkward and intelligent nerd from a conservative upbringing. As a Stanford philosophy student Thiel became a campus provocateur and an adversary of multiculturalism. He was also a closeted gay man who was fiercely protective of his privacy. After graduating from law school in the early 1990s, Thiel was on the East Coast conservative fast track. He clerked for a federal judge, worked as a Manhattan derivatives trader, and then as a Republican speechwriter. Seeking personal fulfillment, however, Thiel returned to Silicon Valley and put his financial and legal talents to work in the dot-com boom.[104]

As Thiel made his fortune, he established himself as a respected innovation expert. His 2011 company memo "What Happened to the Future?" went viral for its call to reform innovation culture. Thiel argued that the venture capital industry had abandoned its commitment to transformational

breakthroughs and instead was chasing "incremental" and "fake problems."[105] He argued that innovation culture needed to return to its "risky,"
"audacious," and "contrarian" roots by tackling grand challenges from
alternative energy to biomedicine and space exploration.

In Thiel's bestselling book, *Zero to One: Notes on Startups, or How to Build
the Future*, the venture capitalist revisited the age-old question of what
made innovators tick. He resurrected Rogers's model of innovators as social
misfits to explain what he called the "founder's paradox." Rather than a
normal distribution, innovators came from bimodal extremes. On one end
were the "weak/nerd," "idiot savant[s]" who were "disagreeable," "poor,"
villainous, and "infamous," and on the other were the charismatic leaders
who were athletic polymaths from elite backgrounds destined to achieved
heroic fame.

Thiel relied on personal experience to make his case. He noted that four
of six of PayPal's founders had "built bombs in high school," and he jok-
ingly asked readers to judge whether his drawing of a hoodie-wearing man
was a tech founder or the Unabomber Ted Kaczynski. "The lesson," Thiel
wrote, was that "we should be more tolerant of founders who seem strange
or extreme" because these "unusual individuals" were necessary to lead
society past "mere incrementalism."[106] Thiel also warned innovators that
the public could turn on them, praising the libertarian novelist Ayn Rand
who imagined a future in which innovators withdrew to a secluded com-
pound to escape a society that persecuted them for their talents.

Thiel concluded that innovators needed to organize rather than retreat.
In the 2012 presidential campaign, he donated $1 million to libertarian
candidate Ron Paul and other fringe politicians with alt-right ties. Then, in
2016, Thiel was one of the few innovation leaders to back Donald Trump.
At the Republican National Convention, he identified himself as a proudly
gay conservative and contrasted "fake culture wars" with a broken and
unequal economy dominated by Silicon Valley. He harkened to a past in
which "the world's high tech capital wasn't just one city: all of America
was high tech" and called for an antiglobalist American nationalism.[107] As
a member of President Trump's transition team, he hoped to dismantle and
remake federal innovation policy. His approach, however, was even more
extreme than the Trump administration's and almost none of his plans
were implemented.[108] In 2020, Thiel expanded his role as a political king-
maker, bankrolling the senate campaigns of his former protege and *Zero*

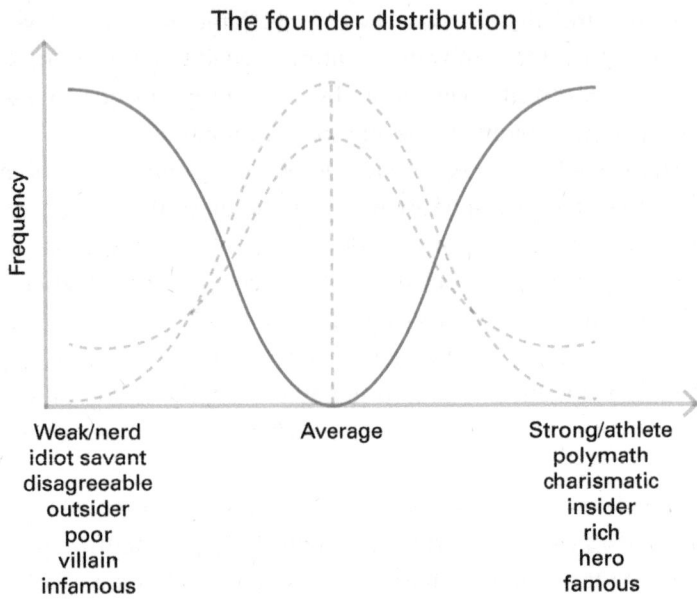

The founder distribution

Frequency

Weak/nerd	Average	Strong/athlete
idiot savant		polymath
disagreeable		charismatic
outsider		insider
poor		rich
villain		hero
infamous		famous

FIGURE 9.3
Innovators as extremists in Peter Thiel's founder distribution. Courtesy of Crown
Publishing Group.

to One coauthor Blake Masters in Arizona and J. D. Vance in Ohio, whose
memoir *Hillbilly Elegy* had popularized the critique of neoliberalism's elit-
ism and inequity.[109]

Where Thiel worked behind the scenes, his former business partner
became the public barometer of the Techlash. In the early 2010s, Musk
appeared to be the next Jobs. He was a living superhero, a boy genius on
whom actor Robert Downey, Jr. modeled his portrayal of *Ironman's* Tony
Stark. A glowing 2015 biography identified Musk as the antidote for the
stagnation of digital culture, a risk-taking genius who plowed his earnings
from PayPal into society's most pressing problems.[110] But as Musk amassed
his fortune and power, he grew more outspoken and erratic. The combination of celebrity romances, a Twitter addiction, and investigative reports
about labor violations and toxic work environments began to tarnish
his image. Online think-pieces on "The Great Elon Musk Debate" asked
whether he was a "hero" or a "villain."[111] Musk's 2018 appearance on The
Joe Rogan Experience podcast, which has been viewed over 66 million times

FIGURE 9.4

In the late 2010s, representations of Elon Musk shifted from hero to antihero. In 2013, *The Atlantic* showcased Musk as the leading edge of technological progress (left). By the end of the decade, public media cast Musk in a darker light as seen in his 2018 appearance on the *Joe Rogan Experience*.

on YouTube alone, captured the tech mogul's evolving reputation. Wearing an "Occupy Mars" T-shirt while smoking a marijuana-laced cigar, Musk explained why his company sold flame throwers, why climate change was an existential problem, and how he was "an alien."[112] Still, in 2021, *Time* named Musk the magazine's Person of the Year. The annual rite emphasized Musk's checkered life and abrasive personality, describing his upbringing by his "evil" father in apartheid South Africa. However, *Time* excused the tech mogul's "earned hubris" for making electric cars a mass-market reality and for reinvigorating the stagnant space industry.[113]

With Musk's takeover of Twitter (which he renamed as "X") the following year, however, he came to epitomize innovation culture's fracture. To conservatives, Musk was a free-thinking heir to Thomas Edison. *National Review* columnist Rich Lowry defended Musk as a counterweight to the conformism of "world famous entrepreneurs [who] are supposed to be all about innovation and disruption, [but] happily let themselves get pulled along in the slipstream of progressive group think."[114] The children's book series Heroes of Liberty placed Musk in a conservative pantheon with Alexander

Hamilton, Harriet Tubman, Ronald Reagan, and Rush Limbaugh. Its book *Elon Musk: Occupy Mars* portrayed the tech billionaire as an exemplar of "the human spirit that reaches for the stars and never wavers." His unreasonableness was the source of his success: "Imagine if people had told him: Grow up! Stop daydreaming! Be realistic! In fact, many did but Elon refused to listen."[115] To liberals, however, Musk embodied innovation extremism. Musk's purchase of Twitter, his barely veiled racism, and his support of free-speech radicalism coincided with his embrace of anti-wokeism. A once star-struck media and entertainment industry turned on Musk, transforming him from Iron Man into a villain who represented the extremism of innovation culture.

Beyond media culture wars, innovation culture's conservatives continue to build a political movement. Following the path blazed by Thiel, in 2023, the entrepreneur and biotechnology investor Vivek Ramaswamy announced his presidential campaign on Fox News' *Tucker Carlson Tonight*. Ramaswamy was a success story from the era of peak innovation. In 2007, he graduated from Harvard and cofounded the Campus Venture Network, a software company that helped student innovators realize their ideas. Two years later, he sold the company to the Kauffman Foundation, a leading innovation philanthropy. He then had a lucrative career in biomedicine. But Ramaswamy came to identify himself as a "class traitor." In his 2021 book, *Woke, Inc: Inside Corporate America's Social Justice Scam*, he argued that progressive politics had infiltrated and corrupted business, and that its assault on meritocracy was stifling innovation.[116] His far-right platform, America First 2.0, argued for shutting down government agencies, pardoning January 6th insurrectionists, abandoning the "climate cult," banning "addictive social media," and ending "unlawful DEI indoctrination."[117] While Ramaswamy's campaign failed, this technonationalist strand of conservative politics made further gains when Trump selected venture capitalist J. D. Vance as his vice-presidential running mate.

Innovation culture's conservatives also developed competing infrastructures for training future innovators. In 2011, Thiel established an alternative to college in the form of a $100,000 grant for "young people who want to build new things instead of sitting in a classroom."[118] Recipients needed to be under twenty-two years old and had to drop out of school to receive funding. Musk meanwhile established the Hyperloop competition, which engaged hundreds of college engineering teams. At the K–12 level, SpaceX developed the Ad Astra school for its employees' children, which grew to become the Synthesis Academy, an online innovator education program for home-schooled children. In many respects, Synthesis Academy resembles the Smithsonian's *Invention at Play*, utilizing images of multicultural teams of "super-collaborators" engaged in creative play. Synthesis Academy's curriculum draws from responsible innovation principles to teach children to "anticipate the consequences of [their] actions by mapping the possible downstream effects." At the same time, however, the school markets itself as a home for outliers. Nine-year-old Ali, for example, "wasn't challenged in traditional schools," but at Synthesis he "found his voice." Fellow nine-year-old Maxwell who has "big dreams," has also "found his people."[119] While these initiatives are small and ad hoc compared to liberal and progressive programs, they reinforced a set of ideals about who innovators should be and how to make them. Like their counterparts, their goal was to sort out the true innovators from the heretics.

SPLIT PERSONALITIES

Reconsider the following profiles of innovators and the means for making them.

In 2022, Ayana Johnson paces the TED stage. Since her 2011 appearance in *Empowering the Nation*, the former NSF grantee has become a climate activist. Blending traditional innovation tropes with their critiques, Johnson's talk encourages people to contribute their "special talents," their "superpowers" to address the climate crisis. She insists that "this is not about being a hero or an influencer," instead, it "is about being useful and advancing solutions." Johnson's initiative, the All We Can Save Project, offers a training program of "wayfinding circles" and "indigenous wisdom" to "hold the powerful to account," by "questioning capitalism," and "standing for justice." Real change, she adds, will only come through antiracist efforts to "demolish the societal barriers that prevent people from fully devoting themselves to climate solutions."[120] Her promise is "a future that holds us, all of us."[121]

Next, let's check back in on Tyson James. *The New Yorker* never followed up on its 2014 satire of the three-year-old disruptor, but we know how the updated satire might read for its audience which has widely turned against innovation culture. James's failed climate venture was a lesson in the hard realities of the market. But James was determined. He went on to win the elementary school division of the Hyperloop contest. He then made a killing in crypto trading. With support from a Thiel Fellowship, he recently dropped out of the eighth grade to work on an AI chatbot that defends against censorship. The latest rumbling is that the teen plans to unseat his octogenarian New Jersey senator, a Big Tech shill.

Although one profile is real and the other imagined, Johnson's and James's life stories underscore the fractured consensus concerning innovation, the divergent characteristics and behaviors of innovators, and the changed perceptions about what innovation means in the twenty-first century.

The imperative to make every American into an innovator reached its apogee in the 2010s. But overpromises and inequities highlighted the pitfalls of innovation culture. In response, innovation experts put forward visions of innovation that were inclusive, responsible, and just. A countermovement portrayed innovation's creative destruction as a necessary price of progress.

What resulted was a fractious view of innovators and innovation culture. The splintering was inextricable with the maelstrom of the Trump years, and in the wake of the Techlash, there are signs of a resurgent innovation imperative. Despite the ongoing demand for innovation, however, its meanings and values are now fundamentally unsettled.

10 EXPERT LESSONS

Everyone is an expert based on their own lived experience.
—Design Justice Network Principles, 2018

We have reached the point where the innovation expert sums up their findings to explain that the secret is creative collaboration, that diversity drives innovation, or that innovators are born outliers. It is alternatively where a different sort of expert reiterates that charlatans have corrupted our institutions and implores us to adopt an alternative mindset.

This book offers neither lesson. Innovation culture has no shortage of evangelists or polemicists. Instead, my aim has been reflective. I have mapped how and why innovation became the dominant global interpretation of science, technology, and society to reconsider the character of the imperative and its possible futures.

The history of innovation culture's rise reveals a series of patterns that have remade life around the world. From the 1950s to today, a remarkably diverse range of people came to believe that innovation has an understandable logic and that innovators equipped with that logic are the key to societal growth.

This imperative grew out of efforts after World War II to answer the biggest questions of a technological age. From where do new things come? What are the sources of wealth? How do ideas travel? How can this combination of factors be harnessed to improve humanity?

Innovation experts were the principal agents behind this vision. In the 1960s and 1970s, these experts constructed a framework for science, technology, and society that challenged the prevailing military–industrial model and its pure science ideal. They extolled innovation's values of purposeful creativity and profitable discovery to meet the challenges of modernity in novel ways. As the number and variety of experts increased in ensuing decades, so did their mindsets and toolkits. Some were charlatans and others were wise beyond their time. Many remain unknown, while a select few gained notoriety as they built their ideas into the world.

At the center of innovation culture was the social type of the innovator. Experts identified the special role of these misfits in generating technological, economic, social, and even moral change. Experts also sought what made innovators tick with the hope of harnessing and reproducing their talents. Extrapolating from their own experiences, these experts identified innovation not as a vocation in the classic sense of science or the professions, nor as a trait with which one was born. Innovation stemmed from a mindset for change and a toolkit for achieving it. Even as innovation experts assumed positions of influence, many perceived *themselves* as upstarts and mavericks, kindred spirits of the innovators that they worked to produce. The mindsets and skillsets that these experts championed applied not just to individuals, but also to organizations, regions, nations, and a global social and economic class.

The extension of innovation expertise from science administrators to elementary-school children democratized how Americans view innovation. Over the decades, who counted as an innovator expanded dramatically across business, engineering, science, the social sciences, humanities, and the arts. So too did the range of methods that they utilized. This pluralism was inherent to the concept of innovation itself. In comparison to *engineering*, *science*, and *technology*, innovation's ambiguity opened the identity of *innovator* to include alternatives aimed at empowering women and minoritized groups. This expansion of innovation expertise was not one of linear diffusion. Instead, innovation's meanings and techniques were appropriated by a wide range of stakeholders from venture capitalists to feminist sociologists and child psychologists. While the most valorized innovators remain business titans like Steve Jobs and Elon Musk, innovation became an inclusive aspiration no matter one's station or life stage.

At the same time the innovation imperative expanded, it narrowed conceptions of what counted as good science, the purposes of art and design, which career paths have value, and what strategies individuals and institutions should pursue. The innovator today presents a personalized image of neoliberal progress. The dominant interpretation of innovation is unremittingly market-driven; even for social innovators, its metric is the financial bottom line.

The innovation imperative was nationalist from the start, channeling visions of the United States' future and one's place within it. Innovation's most vocal champions have been government agents or those funded by or lobbying the government. These bureaucratic innovators pointed internally to domestic needs and externally to the United States' place in the world. Looking inward, research and policy prescriptions applied innovation expertise to seemingly intractable challenges from healthcare to climate change. Looking outward, programs to make innovators emphasized global competitiveness and the fear of a national innovation deficit.

For more than half a century, this ideal of a nation of innovators provided one of the few shared visions in an age of political fracture. Indeed, the bureaucratic apparatus of a national innovation system rose in parallel with the United States' partisan extremism. Projects originally created with liberal motives thrived in conservative regimes and vice versa. This ascendancy was due in part to corporate interests that shaped policy across party lines from the Charpie Report to the CHIPS and Science Act. Throughout innovation conjured a powerful nationalist call for renewal; one that drew on the restoration of the United States' inventive past to tackle growing inequality, centralized military power, environmental destruction, and global competition. This vision of a third way steadily grew across economic booms and busts, balancing hope and anxiety, democracy and technology, talent and inclusion.

In the 2010s, innovation lost its status as an undisputed good. Geographic hubs of innovation culture became symbols of cosmopolitan excess and a cause of rather than a solution to national division. Innovation culture's once creative and rebellious start-ups grew into behemoth corporations with power over the economy, the media, and politics. Profiles of innovators shifted from fawning descriptions of artists who think different to exposés of robber-barons and fascists. Meanwhile, the Trump administration upended the continuity of government initiatives for democratizing innovation, while critics and reformers drew attention to the gulf between innovation culture's rhetoric and its reality.

In as much as this history has a lesson, it is that innovation is neither a passing buzzword nor an inevitable reality; it is an assemblage of ideas and practices built into our technologies, institutions, norms, laws, cultural practices, and inner selves.

The assemblage that we call innovation is at a transformational moment. Globally, the imperative continues to grow along with programs to make innovators of all ages. Hundreds of thousands of people in the United States and millions around the world use the tools of innovation expertise. These techniques, mindsets, and institutions have played key roles in the development of life-saving vaccines and remain crucial for responding to the climate crisis.

At the same time, the world built in innovation's name is replete with overpromises, fraud, and inequity. The dark side of innovation cannot be excised by dismissing bad actors or with platitudes about responsibility. Most innovation experts have either overlooked these negative consequences or promised that their methods are the solution to them.

In response, some critics long for a post-innovation age and argue that we need to abandon the concept. But concepts do not live untethered from the history and culture that give them meaning.[1] Billions of people utilize digital technologies forged in innovation culture that two decades ago did not exist. The innovation imperative has fueled astonishing changes in how we live. Its policy framework is instantiated in infrastructures throughout the world and is ingrained in our most basic visions of self and society. The rapid proliferation of AI portends an acceleration of the trend.

We thus find ourselves again where the first innovation experts started. "We are witnessing changes so profound and far-reaching that the mind can hardly grasp all the implications," John Gardner wrote in 1964. Now, as then, "only the blind and complacent could fail to recognize the great tasks of renewal facing us—in government, in education, in race relations, in urban redevelopment, in international affairs, and most of all in our own minds and hearts."[2] Who, as then, can deny that we require innovation at all levels? However, the imperative has a new refrain for anyone committed to making things better: Innovation by whom? For whom? By what means? And to what ends?

ACKNOWLEDGMENTS

While innovation often rewards moving fast and breaking things, this book has been the result of slow scholarship. Its writing has spanned half a career, a pandemic, and personal and professional milestones. Along the way, I accumulated many debts and I apologize if I've left you out.

I first began thinking about the project for a 2011 Princeton workshop on science and the counterculture organized by David Kaiser and Patrick McCray. Chapter 4, "Be an Innovation Millionaire!" is a revised version of the essay for that event which appeared as "The Birth of Innovation" in *IEEE* Spectrum in February 2015 and "How the Industrial Science Got His Groove" in the edited volume *Groovy Science: Knowledge, Innovation, and American Counterculture* (Chicago 2016), which appear altered with permission from both publishers. I have received feedback on chapters at numerous workshops, conferences, and seminars through the years. Particularly formative were those at the University of Pennsylvania, Johns Hopkins, Technische Universität München, Mines ParisTech, and the University of Toronto's "Techniques of the Corporation" workshop.

Funding from a National Science Foundation Scholar's Award (1354121), a residential fellowship at the Smithsonian Institution's Lemelson Center for the Study of Invention and Innovation, and two Virginia Tech research leaves made the book possible. Any opinions, findings, conclusions, or recommendations expressed are mine and do not necessarily reflect the views of the NSF or any other funder. This support afforded the collection of thousands of documents from the Arthur and Elizabeth Schlesinger Library on the History of Women in America; Carnegie Mellon University Archives;

Gerald R. Ford Presidential Library; Harvard Business School Baker Library; Department of Distinctive Collections, MIT Libraries; National Archives at College Park, MD; Smithsonian Institution Archives; and the Stanford University Special Collections and University Archives. Special thanks to Allison Oswald and Monica Smith at the Lemelson Center for their assistance and institutional memory that made chapter 8 possible. The release time allowed me to sift through thousands of additional resources, especially through the treasure that is Archive.org.

When I started the project, there were almost no critical studies of innovation. I have been lucky to work with and learn from scholars who corrected that deficit. A 2013 unfinished collaboration with the late Benoît Godin—with whom I shared similar questions but different methods—provided the kernel for chapter 2. Visiting positions at MIT's Program in Science, Technology, and Society, the École normale supérieure Paris-Saclay, and the École des hautes études en sciences sociales arranged by David Kaiser, Philippe Fontaine, Aleksandra Kobiljski, Sara Aguiton, and Anne Rasmussen gave me time and space to research, write, and trade ideas. Friends and colleagues who have shared feedback over the years include Ross Bassett, Amy Bix, Jean-Claude Ruano-Borbalan, Dan Bouk, Jonathan Coopersmith, Bretton Fosbrook, Yulia Frumer, Ben Gross, Ryan Hearty, Darshan Karwat, Aisling Kelliher, Brice Laurent, Bill Leslie, Kira Lussier, Joris Mercelis, Cyrus Mody, David Munns, Anderson Norton, Sebastian Pfotenhauer, Amy Slaton, Ellan Spero, Heidi Voskuhl, and Dale Winling.

To understand a culture, you need to experience it; thus, my analysis has drawn as much from lessons learned through critical participation as it has from the archives. As a senior fellow at Virginia Tech's Institute for Creativity, Arts, and Technology (ICAT), I benefited from a long-running conversation with director Ben Knapp about the nature and purposes of transdisciplinary design. My time at ICAT resulted in a series of pedagogical initiatives including the Innovation Pathways Minor, the Human-Centered Design Graduate Education Program, and the Revolutionizing Engineering Departments (NSF 1225856), projects whose joys and frustrations I experienced with Liesl Baum, Ico Bukvic, Luke Lester, Steve Harrison, Tom Martin, Lisa McNair, Dane Webster, and many others. These initiatives not only helped me appreciate the complex motivations of innovation culture's stakeholders, through my course Innovation in Context, it allowed me to share findings and test the book's ideas with hundreds of students, many of whom aspire to "creative class" careers. Participation in the Beyond Boundaries initiative gave me an inside view of the institutional promises and perils of the innovation imperative; I am thankful to Tim Sands, Rosemary Blieszner, Alan Grant, Thanasis Rikakis, and Jill Sible. Finally, at a crucial moment, Rishi Jaitly offered an insider's reflection on the entire manuscript.

A 2014 workshop on innovation expertise that I co-organized with Eric Hintz and Marie Stettler Kleine was a laboratory for comparing the mindsets and toolkits of the champions, critics, and reformers that I discuss in chapter 9. Intended as a field-setting dialogue, the workshop turned out to be a calm before an international storm. It also formed the basis for this book's companion investigation *Does America Need More Innovators?* Thank you to Errol Arkilic, Catherine Ashcraft, W. Bernard Carlson, Lisa D. Cook, Maryann Feldman, Erik Fisher, Benoît Godin, Jenn Gustetic, David Guston, Dutch MacDonald, Sebastian Pfotenhauer, Andrew Russell, Lucinda M. Sanders, Monica Smith, Brenda Trinidad, and Lee Vinsel. Joe Ballay, Bill Lucas, and Mickey McManus at MAYA and LUMA in Pittsburgh and Leticia Britos Cavagnaro and Humera Fasihuddin of the University Innovation Fellows at Stanford's d.school additionally shared their time, insight, and organizational cultures beyond the event. Jeffrey Brodie deserves special thanks for his incredible energy and creativity at the workshop and beyond.

If innovation geographers are to be believed, where we create, and with whom, shapes the character of our ideas. Virginia Tech's STS Department has been an ideal home. I thank every one of my colleagues. Sonja Schmid was a co-conspirator throughout. Saul Halfon, Daniel Breslau, and Rebecca Hester provided aid when I needed it most. Gary Downey reliably answered "yes" whenever I asked if I should again blend the role of historian and practitioner, which seeded the conditions for a growing cluster of critical innovation scholars that includes Lee Vinsel, Fabian Prieto-Ñañez, Fernanda Rosa, and Monamie Haines. Lee and Fabian deserve special commendation for stepping in to teach Innovation in Context over the years.

An amazing group of current and former graduate students provided research assistance or editorial feedback including Monique Dufour, Kendall Giles, Ke Hu, Toma Kawanishi, Foster Oduro Kissi, Sarvnaz Lotfi, Savannah Mandel, Connie McCormack, Michael Meindl, Annie Patrick, Seohyun Park, Desen Özkan, Oliver Shuey, and Danielle Thompson. I am eternally grateful to Kari Zacharias and Marie Stettler Kleine for sorting those thousands of documents, for their brilliant collaborations, and for their concern for my well-being.

I had long believed that a good café is a necessary condition for good writing; 1369, Idego, Rare Bird, and Tandem, thank you. During the pandemic, however, writing collectives replaced the coffee house. Paola Zellner, Greg Galford, and the rest of the COAST Writing Group have been ideal companions for twice weekly sessions of incremental progress. Danna Agmon's Wednesday Writers group has also been a sanctuary.

Thank you to the MIT Press publishing team of Katie Helke, Justin Kehoe, Suraiya Jetha, Pamela Quick, and Roger Wood. Arthur Daemmrich and Art Molella, both past Lemelson Center directors, have been encouraging and generous series editors. Anonymous reviewers helped me make crucial improvements. Audra Wolfe and Kathleen Kearns offered editorial advice at a few key moments. Jitendra Kumar of Westchester Publishing Services has guided me through the book's production. The Faculty Subvention Fund at Virginia Tech supported the index by Sanjiv Kumar Sinha. Corinne Guimont and the TOME program deserve the thanks of everyone reading this for making the digital edition free and open to all.

My family's love has kept the project going through good and bad. Isaak and Aaron, who have no memory of a time when I haven't been working on the book, are now encountering the world it describes. Cindy Rosenbaum, for whom this project is a blink in the eye of our forever partnership, is surely happy it is done. My second Blacksburg family Chris Williams and Justine Brantley provided warm fires, amazing barbeque, and *checks notes* peanut butter whiskey during our shared quarantine. Thank you to my parents and siblings who opened their homes on three continents and knew when and when not to ask about innovation.

Finally, I thank my series editors Eric Hintz and Joyce Bedi. Eric is a walking encyclopedia of the history of invention and innovation. More important, he is a generous soul who made every page better. In March 2013, Joyce Bedi approached me after a talk to say she'd like to help me spread my ideas. From the very start, she gave me the gift of confidence. She has delivered that gift again and again. This book is for Joyce; its story is hers.

NOTES

PREFACE

1. For details on these critical participation projects, see Matthew Wisnioski and Kari Zacharias, "Sandbox Infrastructure: Field Notes from the Arts Research Boom," *ARPA Journal* 1, no. 1 (2014), http://arpajournal.gsapp.org/we-are-test-subjects-2; Kari Zacharias and Matthew Wisnioski, "Land-Grant Hybrids: From Art and Technology to SEAD," *Leonardo* 52, no. 3 (2019): 275–284; Annie Y. Patrick et al., "In it for the Long Haul: The Groundwork of Interdisciplinary Culture Change in Engineering Education Reform," *Engineering Studies*, 15, no. 2 (2023): 144–167; Rosemary Blieszner, Alan Grant, and Thanasis Rikakis, *Envisioning Virginia Tech Beyond Boundaries: A 2047 Vision* (Blacksburg, VA: Virginia Tech, 2016), https://beyondboundaries.vt.edu/assets/visioning-document-print.pdf; Matthew Wisnioski and Lee Vinsel, "The Campus Innovation Myth: A Half-Century of Occasional Breakthroughs and Many Disappointments," *Chronicle of Higher Education* (June 21, 2020), B6–B9; Matthew Wisnioski, Eric Hintz, and Marie Stettler Kleine, eds., *Does America Need More Innovators?* (Cambridge, MA: MIT Press, 2019).

CHAPTER 1

1. The Pittsburgh Foundation, "Change Agents for Education," February 13, 2014, Vimeo video, https://vimeo.com/76618571; LUMA, "LUMA HCD Innovation Camp," October 1, 2009, YouTube video, https://www.youtube.com/watch?v=XNJ69PXB9Ww; "Change Agents in Education," The Pittsburgh Foundation, https://pittsburghfoundation.org/change-agents.

2. "LUMA Institute," https://www.luma-institute.com; Kitty Julian, "Becoming the Change," *Pittsburgh Foundation Quarterly* (Winter 2016): 6–7.

3. Langdon Winner, "The Cult of Innovation: Its Colorful Myths and Rituals," *Langdon Winner on Politics, Technology and the Arts*, June 12, 2017, https://www .langdonwinner.com/other-writings/2017/6/12/the-cult-of-innovation-its-colorful -myths-and-rituals; Lee Vinsel, "Design Thinking Is Kind of Like Syphilis: It's Contagious and Rots Your Brains," December 6, 2017, https://medium.com/@sts _news/design-thinking-is-kind-of-like-syphilis-its-contagious-and-rots-your-brains -842ed078af29.

4. Benoît Godin, *Innovation Contested: The Idea of Innovation Over the Centuries* (London: Routledge, 2015).

5. Sebastian Pfotenhauer and Sheila Jasanoff, "Panacea or Diagnosis? Imaginaries of Innovation and the 'MIT Model' in Three Political Cultures," *Social Studies of Science* 47, no. 6 (2017): 783–810.

6. Mike Ramsey and Douglas MacMillan, "Carnegie Mellon Reels After Uber Lures Away Researchers," *Wall Street Journal*, May 31, 2015, https://www.wsj.com/articles /is-uber-a-friend-or-foe-of-carnegie-mellon-in-robotics-1433084582.

7. Pittsburgh Technology Council, "2017–2018 Pittsburgh Techmap," 2018, https://web.archive.org/web/20191121054148/https://www.pghtech.org/2017-18 -pittsburgh-techmap.aspx.

8. Megan Trotter, "A Stroll Down 'AI Avenue': Pittsburgh's New Tech Hub," *Trib Live*, June 21, 2024, https://triblive.com/business/technology/a-stroll-down-ai-avenue -pittsburghs-new-tech-hub/

9. Matthew Wisnioski, *Engineers for Change: Competing Visions of Technology in 1960s America* (Cambridge, MA: MIT Press, 2012); Eric Schatzberg, *Technology: Critical History of a Concept* (Chicago, IL: University of Chicago Press, 2018).

10. Henry Etzkowitz, *The Triple Helix: University-Industry-Government Innovation in Action* (New York: Routledge, 2008); Michael Gibbons, *The New Production of Knowledge: The Dynamics of Science and Research in Contemporary Societies* (London: SAGE Publications, 1994).

11. Maryann P. Feldman, *The Geography of Innovation: Economics of Science, Technology, and Innovation* (Dordrecht and Boston: Kluwer Academic, 1994); AnnaLee Saxenian, *Regional Advantage: Culture and Competition in Silicon Valley and Route 128* (Cambridge, MA: Harvard University Press, 1994).

12. Walter Isaacson, *Steve Jobs* (New York: Simon & Schuster, 2011).

13. Clayton M. Christensen, *The Innovator's Dilemma: When New Technologies Cause Great Firms to Fail* (Boston, MA: Harvard Business School Press, 1997); Richard L. Florida, *The Rise of the Creative Class, and How It's Transforming Work, Leisure, Community, and Everyday Life* (New York: Basic Books, 2002).

14. MAYA stands for "Most Advanced Yet Acceptable," a mantra of industrial design founder Raymond Loewy.

15. Tom Kelley and David Kelley, *Creative Confidence: Unleashing the Creative Potential Within Us All* (New York: Crown, 2013).

16. Lee Vinsel and Andrew L. Russell, *The Innovation Delusion: How Our Obsession with the New Has Disrupted the Work That Matters Most* (New York: Currency, 2020).

17. Allen Dieterich-Ward, *Beyond Rust: Metropolitan Pittsburgh and the Fate of Industrial America* (Philadelphia, PA: University of Pennsylvania Press, 2016).

18. *Innovative Regions: The Importance of Place and Networks in the Innovative Economy* (Palo Alto, CA: Collaborative Economics, 1999).

19. Luke Torrance, "Report: Pittsburgh Named One of Most Livable Cities in U.S.," *Pittsburgh Business Journal*, September 5, 2019, https://www.bizjournals.com /pittsburgh/news/2019/09/05/reportpittsburgh-named-one-of-most-livable-cities .html.

20. Thomas L. Friedman, *The World is Flat: A Brief History of the Twenty-First Century* (New York: Farrar, Straus and Giroux, 2005); National Academy of Sciences, National Academy of Engineering, and Institute of Medicine, *Rising Above the Gathering Storm: Energizing and Employing America for a Brighter Economic Future* (Washington, DC: National Academies Press, 2007).

21. Stuart W. Leslie, "Regional Disadvantage: Replicating Silicon Valley in New York's Capital Region," *Technology and Culture* 42, no. 2 (2001): 236–264.

22. Helga Nowotny, *Insatiable Curiosity: Innovation in a Fragile Future* (Cambridge, MA: MIT Press, 2008), 8.

23. David Harvey, *A Brief History of Neoliberalism* (New York: Oxford University Press, 2005).

24. My analysis is influenced by Steven Shapin's notion of the scientific self. Steven Shapin, *The Scientific Life: A Moral History of a Late Modern Vocation* (Chicago, IL: University of Chicago Press, 2008).

CHAPTER 2

1. Everett M. Rogers, "A Conceptual Variable Analysis of Technological Change," (PhD diss. Iowa State College, 1957).

2. F. L. Timmons, "A History of Weed Control in the United States and Canada," *Weed Science* 18, no. 2 (1970): 294–307.

3. Everett M. Rogers, *Diffusion of Innovations* (New York: The Free Press, 1962); Elliott Green, "What Are the Most-cited Publications in the Social Sciences (According to Google Scholar)?" *LSE Impact Blog*, May 12, 2016, https://blogs.lse.ac.uk /impactofsocialsciences/2016/05/12/what-are-the-most-cited-publications-in-the -social-sciences-according-to-google-scholar/. The most cited social science title is Thomas Kuhn, *The Structure of Scientific Revolutions* (Chicago, IL: University of

Chicago Press, 1962). Kuhn, like Rogers, presents a theory, the paradigm shift, for how new ideas are adopted.

4. Benoît Godin, "How Innovation Evolved from a Heroic Act to a Heroic Imperative," in *Does America Need More Innovators?* ed. Matthew Wisnioski, Eric S. Hintz, and Marie Stettler Kleine, eds. (Cambridge, MA: MIT Press, 2019), 141–164.

5. Gabriel Tarde and Elsie Worthington Clews Parsons, *The Laws of Imitation* (New York: H. Holt and Company, 1903).

6. William Fielding Ogburn, "The Influence of Invention and Discovery," in *Recent Social Trends in the United States* (New York: McGraw-Hill Book Company, Inc., 1933); Benoît Godin, *Models of Innovation: The History of an Idea* (Cambridge, MA: MIT Press, 2017), 18–21.

7. Margaret Mead, *And Keep Your Powder Dry: An Anthropologist Looks at America* (New York: Morrow, 1965).

8. Stuart W. Leslie, *The Cold War and American Science: The Military-Industrial-Academic Complex at MIT and Stanford* (New York: Columbia University Press, 1993); Godin, *Models of Innovation*.

9. Sarvnaz Lotfi, "Capitalizing the Measure of Our Ignorance: A Pragmatist Genealogy of RandD," (PhD diss. Virginia Polytechnic Institute and State University, 2020).

10. David A. Hounshell, "The Evolution of Industrial Research in the United States," in *Engines of Innovation: U.S. Industrial Research at the End of an Era*, ed. Richard S. Rosenbloom and William J. Spencer (Boston, MA: Harvard Business School Press, 1996); Leonard S. Reich, *The Making of American Industrial Research: Science and Business at GE and Bell, 1876–1926* (Cambridge, UK: Cambridge University Press, 1985).

11. Jon Gertner, *The Idea Factory: Bell Labs and the Great Age of American Innovation* (New York: Penguin, 2012).

12. Martin Watzinger, Thomas A. Fackler, Markus Nagler, and Monika Schnitzer, "How Antitrust Enforcement Can Spur Innovation: Bell Labs and the 1956 Consent Decree," *American Economic Journal: Economic Policy* 12, no. 4 (2020): 328–359.

13. "Laboratories Marks a Decade of Transistor Progress," *Bell Laboratories Record* 46, no. 8 (1958): 304–306; Gertner, *The Idea Factory*, fn1.

14. Bell Telephone Laboratories, H. E. Bridgers, and F. J. Biondi, *Transistor Technology*, 3 vols., Bell Telephone Laboratories series, (Princeton, NJ: Van Nostrand, 1958).

15. John R. Pierce, "Innovation in Technology," *Scientific American* 199, no. 3 (1958), 116–134.

16. W. Bernard Carlson, "Innovation and the Modern Corporation: From Heroic Inventor to Industrial Science," in *Science in the Twentieth Century*, ed. John Krige and Dominque Pestre (Amsterdam, Netherlands: Harwood, 1997), 203–226.

17. Eric S. Hintz, *American Independent Inventors in an Era of Corporate R&D* (Cambridge, MA: MIT Press, 2021).

18. Kenneth C. E. Mees, "The Organization of Industrial Scientific Research," *Science* 43, no. 1118 (1916): 763–773. The distinction, however, served to highlight the shifting fortunes of the two related concepts before and after World War II. In Joseph Rossman's 1933 study *The Psychology of the Inventor*, a cherished text of inventor advocates, the term "innovation" appeared on just 4 pages compared to 193 that mention "invention." Joseph Rossman *The Psychology of the Inventor: A Study of the Patenee* (Washington, DC: The Inventors Publishing Co. 1931).

19. On the role of government funding in postwar R&D, see Stuart W. Leslie, "The Biggest 'Angel' of Them All: The Military and the Making of Silicon Valley," in *Understanding Silicon Valley: The Anatomy of an Entrepreneurial Region*, ed. Martin Kenney (Stanford, CA: Stanford University Press, 2000), 48–68.

20. Donald A. Schön, "Champions for Radical New Inventions," *Harvard Business Review*, 41 (1963): 77–86.

21. Homer Garner Barnett, *Innovation: The Basis of Cultural Change* (New York: McGraw-Hill Book Company, Inc., 1953).

22. For a summary see Hornell Hart, "Some Cultural-Lag Problems Which Social Science Has Solved," *American Sociological Review* 16, no. 2 (1951): 223–227; Benoît Godin, "Innovation Without the Word: William F. Ogburn's Contribution to the Study of Technological Innovation," *Minerva* 48, no. 3 (2010): 277–307.

23. Elting E. Morison, "A Case Study of Innovation," *Engineering and Science Monthly* 13, no. 7 (April 1950): 5–11.

24. Philip Lowe, "Enacting Rural Sociology: Or what are the Creativity Claims of the Engaged Sciences?" *Sociologia Ruralis* 50, no. 4 (2010): 313–314.

25. A. Lee Coleman et al., *How Farm People Accept New Ideas* (Ames, IA: Agricultural Extension Service Iowa State College, 1955).

26. Bryce Ryan and Neal C. Gross, "The Diffusion of Hybrid Seed Corn in Two Iowa Communities," *Rural Sociology* 8, no. 1 (1943): 15–24.

27. William L. Brown, "H. A. Wallace and the Development of Hybrid Corn," *The Annals of Iowa* 47, no. 2 (1983): 167–179.

28. Deborah Fitzgerald, *Every Farm a Factory: The Industrial Ideal in American Agriculture* (New Haven, CT: Yale University Press, 2010).

29. Ryan and Gross, "The Diffusion of Hybrid Seed Corn in Two Iowa Communities," 24.

30. Lowe, "Enacting Rural Sociology," 319; Irwin T Sanders, *Interprofessional Training Goals for Technical Assistance Personnel Abroad: Report of an Interprofessional Conference on Training of Personnel for Overseas Service*, Cornell University (Ithaca, NY: Council

on Social Work Education, 1959). In the 1950s, there were also parallel developments in a European context. See, for example, Torsten Hägerstrand, *Innovation Diffusion as a Spatial Process* (Chicago, IL: University of Chicago Press, 1967).

31. Thomas W. Valente and Everett M. Rogers, "The Origins and Development of the Diffusion of Innovations Paradigm as an Example of Scientific Growth," *Science Communication* 16, no. 3 (1995): 252–257. On the report's spread see Rogers, *Diffusion of Innovations*, 62.

32. James Coleman, Elihu Katz, and Herbert Menzel, "The Diffusion of an Innovation Among Physicians," *Sociometry* 20, no. 4 (1957), 253–270.

33. Joseph A. Schumpeter, "The Analysis of Economic Change," *The Review of Economics and Statistics* 17, no. 4 (1935): 2–10.

34. Thomas K. McCraw, *Prophet of Innovation: Joseph Schumpeter and Creative Destruction* (Cambridge, MA: Belknap Press, 2010); Arnold Heertje, "Schumpeter and Technical Change," in *Evolutionary Economics: Applications of Schumpeter's Ideas*, ed. Horst Hanusch (Cambridge, UK: Cambridge University Press, 1988); John Hagedoorn, "Innovation and Entrepreneurship: Schumpeter Revisited," *Industrial and Corporate Change* 5, no. 3 (1996): 883–896.

35. Joseph A. Schumpeter, *Capitalism, Socialism and Democracy* (New York: Harper and Brothers, 1942).

36. When describing his discovery, Solow quipped that "One notes with satisfaction that the trend is strongly upward; had it turned out otherwise I would not now be writing this paper." Robert M. Solow, "Technical Change and the Aggregate Product Function," *The Review of Economics and Statistics* 39, no. 3 (1957): 312–320.

37. Verena Halsmayer, "From Exploratory Modeling to Technical Expertise: Solow's Growth Model as a Multipurpose Design," *History of Political Economy* 46, no. 1 (2014): 229–251.

38. Zvi Griliches, "Hybrid Corn: An Exploration in the Economics of Technological Change," *Econometrica* 25, no. 4 (1957): 501–522. Griliches had been a student of agricultural economist Theodore Schultz, who once worked at Iowa State's agricultural experiment station. Arthur Aaron Bright, *The Electric-lamp Industry: Technological Change and Economic Development from 1800 to 1947* (New York: Macmillan, 1949); Yale Brozen, "Invention, Innovation, and Imitation," *The American Economic Review* 41, no. 2 (1951): 239–257.

39. W. Rupert Maclaurin, "The Sequence from Invention and its Relation to Economic Growth," *The Quarterly Journal of Economics* 67, no. 1 (1953): 97–111.

40. David A. Hounshell, "The Medium is the Message, or How Context Matters: The RAND Corporation Builds an Economics of Innovation, 1946–1962," in *Systems, Experts, and Computers: The Systems Approach in Management and Engineering, World War II and After*, ed. Agatha C. Hughes and Thomas P. Hughes (Cambridge, MA: MIT

Press, 2000); National Bureau of Economic Research, *The Rate and Direction of Inventive Activity: Economic and Social Factors* (Princeton, NJ: Princeton University Press, 1962); Josh Lerner and Scott Stern, *The Rate and Direction of Inventive Activity Revisited* (Chicago, IL: University of Chicago Press, 2012).

41. Melissa Banta, "Georges F. Doriot: Educating Leaders, Building Companies," Harvard Business School, 2014, https://www.library.hbs.edu/hc/doriot/; Spencer E. Ante, *Creative Capital: Georges Doriot and the Birth of Venture Capital* (Cambridge, MA: Harvard Buisness Review Press, 2008).

42. David H. Hsu and Martin Kenney, "Organizing Venture Capital: the Rise and Demise of American Research and Development Corporation, 1946–1973," *Industrial and Corporate Change* 14, no. 4 (2005): 579–616.; Paul A. Gompers, "The Rise and Fall of Venture Capital," *Business and Economic History* 23, no. 2 (1994): 1–24.

43. Georges Doriot, Notes for Affiliate Dinner, 1962, April 17, 1973. Georges Doriot Papers, Baker Library, Harvard Business School.

44. Electronics Seminar, Blair & Co., Chicago, "Electronics Revolution" and "Venture Capital," March 7, 1967, Georges Doriot Papers, Baker Library, Harvard Business School.

45. Art Kleiner, *The Age of Heretics: Heroes, Outlaws, and the Forerunners of Corporate Change*, 1st ed. (New York: Currency Doubleday, 1996); Javier Lezaun, "Demo for Democracy," *Limn*, Issue 0, https://limn.it/articles/demo-for-democracy/.

46. Management theorists such as Chris Argyris bolstered T-group techniques with research on the stifling effect of corporate bureaucracies. Chris Argyris, *Personality and Organization: The Conflict Between System and the Individual* (New York: Harper, 1957).

47. Ronald Lippitt, *The Dynamics of Planned Change: A Comparative Study of Principles and Techniques* (New York: Harcourt, 1958), 4–5.

48. Bennis had no advanced mathematics training and reminisces about being completely lost in a class he took with Solow. Warren G. Bennis, *An Invented Life: Reflections on Leadership and Change* (New York: Basic Books, 1994), 16–20.

49. Warren G. Bennis, "The Social Science Research Organization: A Study of the Institutional Practices and Values in Interdisciplinary Research," (PhD Diss. Massachusetts Institute of Technology, 1954).

50. Warren G. Bennis, *The Planning of Change: Readings in the Applied Behavioral Sciences* (New York: Holt, 1961), 6; Kleiner, *The Age of Heretics*, 236–239.

51. John W. Gardner, *Excellence, Can We Be Equal and Excellent Too?* (New York: Harper, 1961).

52. John W. Gardner, *Self-Renewal: The Individual and the Innovative Society* (New York: Harper & Row, 1964), 127.

53. While collaboration between government and business have always existed, the language and theorization of public–private partnerships was relatively novel to the period. Julie L. Plaut, "'The Citizen of His Era': John W. Gardner and Public Life in the Twentieth-Century United States," (PhD diss. Indiana University, 2003), 82–96.

54. Gardner, *Excellence*, 86.

55. Rogers, *Diffusion of Innovations*, 17.

56. Thomas E. Backer et al., "Writing with Ev: Words to Transform Science into Action," *Journal of Health Communication* 10 (2005): 294. Everett M. Rogers et al., *The Fourteenth Paw: Growing Up on an Iowa Farm in the 1930s* (Singapore: Asian Media Information and Communication Centre and Wee Kim Wee School of Communication and Information, Nanyang Technological University, 2008).

57. Arvind Singhal and James W. Dearing, ed., *Communication of Innovations: A Journey with Ev Rogers* (New Delhi, India: SAGE Publications India Pvt Ltd, 2006), 20.

58. Arvind Singhal, "Everett M. Rogers, an Intercultural Life: From Iowa Farm Boy to Global Intellectual," *International Journal of Intercultural Relations* 36 (2012): 850.

59. Everett M. Rogers, "Diffusion of Human Factors Design: Resistances and How to Overcome Them," *Proceedings of the Human Factors and Ergonomics Society Annual Meeting* 41, no. 1 (2016): 1–3; "Pilots' Pants Problem Proves Problem for Professor," *Logan Daily News*, January 29, 1959, 2.

60. Singhal, "Everett M. Rogers, an Intercultural Life," 850.

61. Rogers, *Diffusion of Innovations*, 159. See also George M. Beal and Everett M. Rogers, *The Adoption of Two Farm Practices in a Central Iowa Community* (Ames, IA: Agricultural and Home Economics Experiment Station, Iowa State University of Science and Technology, 1960).

62. Rogers, *Diffusion of Innovations*, 279.

63. Rogers, *Diffusion of Innovations*, 60.

64. Traditional norms were not "necessarily" undesirable because they provided stability during moments of rapid societal change. Traditional social systems also had redeeming qualities, such as valuing "friendliness and hospitality." Rogers, *Diffusion of Innovations*, 61–62.

65. Edwin B. Parker, "Review: Studies in Diffusion," *AV Communication Review* 12, no. 2 (1964): 221–222; B. F. Stanton, "Review: Diffusion of Innovations," *Journal of Farm Economics* 45, no. 4 (1963): 898; Mario J. A. Bick, "Review: Diffusion of Innovations," *American Anthropologist* 65, no. 5 (1963): 1146–1147.

66. William W. Wayson, "Review: Diffusion of Innovations," *Theory into Practice* 2, no. 5 (1963): 287–290.

67. Everett M. Rogers, "A Prospective and Retrospective Look at the Diffusion Model," *Journal of Health Communication* 9, Suppl 1 (2004): 17. Today, the book

is in its fifth edition, and innovation studies number in the tens of thousands. In 1973, Rogers also identified 500 diffusion studies just on family planning in India. Everett M. Rogers, *Communication Strategies for Family Planning* (New York: Free Press, 1973).

68. For a partial record of Rogers's numerous speeches, see "Extension Service Report for 1960," *Washington C.H. Record*, March 4, 1960, 2; "Institute Opens Today At Hardin-Northern School," *The Republican Courier*, March 9, 1961, 14; "Junior Leaders, Delegates to Hear Dr. Everett Rogers," *Lancaster Eagle Gazette*, August 31, 1962, 12.

69. Everett M. Rogers, *Communication and Development: Critical Perspectives* (Beverly Hills: SAGE Publications, 1976).

70. North Central Rural Sociology Committee, *Adopters of New Farm Ideas: Characteristics and Communications Behavior* (Lansing, MI: North Central Rural Sociology Committee, 1961).

71. Eugene A. Wilkening, "Informal leaders and innovators in farm practices," *Rural Sociology* 17, no. 3 (1952): 272–275.

72. Experts used varying labels to describe the range of human responses to innovation. The earliest studies by Tarde drew distinctions between innovators and imitators. Tarde and Parsons, *The Laws of Imitation*. Economic studies in the 1940s used four categories, although they defined innovators as firms rather than individuals. Terms used in the 1950s to describe innovators included "advanced scouts," "lighthouses," "earliest acceptors," "non-parochials," the "cultural *avant garde*," and "progressists." "Early adopters" were sometimes referred to as "spark plugs," "technical leaders," "opinion leaders," or just "leaders." The least likely to adopt an innovation were "laggards," "parochials," "diehards," "traditionalists," and "drones." Rogers, *Diffusion of Innovations*, 148, 150–151.

73. Rogers, *Diffusion of Innovations*, 207.

74. Francis Stuart Chapin, *Cultural Change* (New York: The Century Co., 1928), 394.

75. Barnett, *Innovation*, 380.

76. Rogers and Havens identified these qualities by interviewing 99 Ohio farmers identified as innovators by state extension agents. An initial random sample of 104 farmers turned up just three innovators, confirming Rogers's belief in a distribution of 2.5 percent. Everett M. Rogers, "What Are Innovators Like?" *Theory into Practice* 2, no. 5 (1963): 252–256.

77. Rogers, "What Are Innovators Like?" 253–254.

78. Rogers, *Diffusion of Innovations*, 169.

79. Edward B. Roberts, "A Basic Study of Innovators: How to Keep and Capitalize on Their Talents," *Research Management* 11, no. 4 (1968): 249–266.

80. G. Gordon and S. Marquis, "Freedom, Visibility of Consequences, and Scientific Innovation," *American Journal of Sociology* 72, no. 2 (1966): 195–202.

81. Thomas S. Robertson, *Innovative Behavior and Communication* (New York: Holt, 1971), 84–118.

82. Banesh Hoffmann, "The Influence of Albert Einstein," *Scientific American* 180, no. 3 (1949): 53. Among the article's motives was to dispel the stereotype of scientists were elitist intellectuals.

83. Robert Squeri, "Leonardo de Vinci: Innovator," *Art Education* 14, no. 9 (December 1961): 6–9, 15.

84. "Executives Lean to Organization Men, Not Innovators, in Picking Aides," *BusinessWeek*, September 23, 1961, 168; Richard Samuel Reynolds, "Lessons of Leadership: Being an Innovator," *Nation's Business*, February 1966, 66.

85. Spencer Klaw, "The Cultural Innovators," *Fortune* 62 (February 1960): 147–149, 202, 207–208, 211.

86. Matthew Josephson, *Edison: A Biography*, 1st ed. (New York: McGraw-Hill, 1959), xiii–xiv, 261–64. For the history of valorizations of inventors, see Louis C. Hunter, "The Heroic Theory of Invention," in *Technology and Social Change in America*, ed. Jr. Layton, Edwin T. (New York: Harper & Row, 1973), 25–46.

87. Baldwin Ward, *The 50 Great Pioneers of American Industry: The Stories of Rockefeller, Swift, Edison, Woolworth, Squibb, Proctor, Sears, Otis, Singer, Carrier, and 40 other Business Leaders and Courageous Innovators Whose Activities Founded Major Industries and Shaped Today's Economy* (Chicago, IL: J.G. Ferguson, 1965), 4.

88. Papers surveyed include *New York Times, Washington Post, Los Angeles Times, Chicago Tribune, Wall Street Journal*, and *Atlanta Constitution*. The first such ad appeared in 1950, followed by 10 more total in the period 1950–1959. Between 1960 and 1964, the number rose to 64 ads. Between 1965 and 1969, 1,051 ads.

89. Katherine E. Hill, "The Recognition And Significance Of Children As Innovators," *Science and Children* 5, no. 1 (September 1967): 29–30, 32–33.

90. Rogers, *Diffusion of Innovations*, 2, 143.

91. Barnett, *Innovation*, 35.

92. Gardner, *Self-Renewal*, 30–31.

CHAPTER 3

1. Daniel V. De Simone, Fact Sheet and Press Release: Innovation Prizes [Draft], September 11, 1972, RG 359 Records of the Office of Science and Technology Policy, Records of the Staff Handling the Presidential Prize for Innovation (PPI-72), Miscellaneous Staff Files, box 3, National Archives at College Park, MD (hereafter cited as PPI OSTP 359).

2. De Simone, Fact Sheet and Press Release: Innovation Prizes [Draft].

3. Deborah Shapley, "The Presidential Prize Caper," *Science* 183, no. 4128 (1974): 938.

4. The chapter builds on arguments in Fred L. Bock, *States of Innovation: The U.S. Government's Role in Technology Development* (Milton Park, UK: Routledge, 2011) and Mariana Mazzucato, *The Entrepreneurial State: Debunking Public vs. Private Sector Myths* (New York: Anthem Press, 2013). Here, and to a lesser extent in chapter 6, I recount developments also explored by Benoît Godin in *The Idea of Technological Innovation: A Brief Alternative History* (Cheltenham, UK: Edward Elgar, 2020), which I discovered (and which was published) after I completed the research and writing for this chapter.

5. Richard R. Nelson, *National Innovation Systems: A Comparative Analysis* (New York: Oxford University Press, 1993); Bengt-Ake Lundvall, *National Systems of Innovation: Towards a Theory of Innovation and Interactive Learning* (New York, London: Pinter Publishers, 1992).

6. Most scholars link the government's role in innovation to the Bayh–Dole Act of 1980. Important exceptions are Elizabeth Popp Berman, *Creating the Market University: How Academic Science Became an Economic Engine* (Princeton, NJ: Princeton University Press, 2012), 40–49 and John D. Skrentny and Natalie Novick, "From Science to Alchemy: The Progressives' Deployment of Expertise and the Contemporary Faith in Science to Grow the Economy and Create Jobs," in *The Progressives' Century: Political Reform, Constitutional Government, and the Modern American State*, ed. Stephen Skowronek, Stephen M. Engel, and Bruce Ackerman (New Haven, CT: Yale University Press, 2016), 405–427.

7. Benoît Godin, "The Linear Model of Innovation: The Historical Construction of an Analytical Framework," *Science, Technology & Human Values* 31, no. 6 (2006): 639–667.

8. Hollomon's presidential level appointment resulted from a National Academy of Sciences committee led by Mervin Kelly, the former director of Bell Laboratories. National Academy of Sciences, *The Role of the Department of Commerce in Science and Technology* (Washington, DC: National Academy of Sciences, 1960).

9. Helen Bowers, *From Lighthouses to Laserbeams: A History of the U.S. Department of Commerce* (Washington, DC: U.S. Department of Commerce, Office of the Secretary, 1995).

10. In 1963, President Kennedy built on Hollomon's talking points to describe "a wealth of new opportunities for innovation," but also how the nation had "paid a price by sharply limiting the scarce scientific and engineering resources available to the civilian sectors of the American economy." John F. Kennedy, "Excerpts From Annual Message to the Congress: The Economic Report of the President," Online by Gerhard Peters and John T. Woolley, The American Presidency Project https://www

.presidency.ucsb.edu/node/237159. For the influence of Hollomon on this speech, see J. Herbert Hollomon, "Proposed Special Message of the President on Technology, Economic Prosperity, and National Security," January 16, 1963, RG 40, General Records of the Department of Commerce, box 60, Hollomon, National Archives at College Park, MD (hereafter cited as Commerce RG 40).

11. Memorandum from J. Herbert Hollomon to Daniel V. De Simone, "Operational Home for the Office of Invention and Innovation," May 26, 1964, Scientific and Technological Files, Office of the Secretary Office of Science and Technology, Commerce RG 40.

12. Hollomon also compared US productivity unfavorably to growth rates in Japan, West Germany, the Netherlands, Sweden, the UK, Canada, and the Soviet Union. J. Herbert Hollomon, "Science, Technology, and Economic Growth," *Physics Today* 16 (1962): 38–42.

13. "The Institute for Applied Technology," Assistant Secretary for Science and Technology, box 12, Institute for Applied Technology, Commerce RG 40.

14. Summary of Minutes of National Inventors Council, September 5, 1963, Assistant Secretary for Science and Technology, box 25, National Inventors Council, Commerce RG 40.

15. Dorothy Nelkin, *The Politics of Housing Innovation: The Fate of the Civilian Industrial Technology Program* (Ithaca, NY: Cornell University Press, 1971). Myron Tribus to Daniel V, De Simone from Roger Gilbertson, "Postmortem on STS," January 28, 1970, Subject File 1962–1970, Office of the Secretary Office of Science and Technology, Commerce RG 40.

16. Office of State Technical Services, *New Technology and the American Economy: A Report of the Public Evaluation Committee* (Washington, DC: United States Department of Commerce, 1968).

17. "U.S. Promotes Spread of Technical Innovation," in *CQ Almanac 1965* (Washington, DC: Congressional Quarterly, 1966), 918–921.

18. Lyndon B. Johnson, "Remarks at the Signing of the State Technical Services Act," Online by Gerhard Peters and John T. Woolley, The American Presidency Project https://www.presidency.ucsb.edu/node/240560.

19. Joseph A. Califano, Jr., "The Politics of Innovation and the Revolution in Government Management," in *Science and Technology Fellowship Program, 1966–1967* (Washington, DC: Department of Commerce, 1966), 27–42.

20. Lyndon B. Johnson, Annual Message to the Congress on the State of the Union. Online by Gerhard Peters and John T. Woolley, The American Presidency Project https://www.presidency.ucsb.edu/node/238437. For more on creative federalism, see Bell Julian Clement, "Creative Federalism and Urban Policy: Placing the City in

the Great Society," (PhD diss. George Washington University, 2014); Richard Ralph Warner, "The Concept of Creative Federalism in The Johnson Administration," (PhD diss. American University, 1970).

21. Memorandum from Hollomon to De Simone, "Operational Home for the Office of Invention and Innovation," May 26, 1964, Scientific and Technological Files, Office of the Secretary Office of Science and Technology, Commerce RG 40.

22. Arthur D. Little Inc., *Patterns and Problems of Technical Innovation in American Industry* (Washington, DC: US Government Printing Office, 1963).

23. Daniel V. De Simone, "Innovation and the Unreasonable Man," 1965, Dep. Dir-Experiment R&D Inc. Files, box 1, Papers from Non-Federal Sources, NSF RG 307; Daniel V. De Simone, "Invention, the First Step, Is Often Most Difficult," *New York Times*, January 8, 1968, 138.

24. US Department of Commerce, *Technological Innovation: Its Environment and Management* (Washington, DC: US Government Printing Office, 1967), 2.

25. US Department of Commerce, *Technological Innovation*, 2.

26. US Department of Commerce, *Technological Innovation*, 56.

27. Sarah Bridger, *Scientists at War: The Ethics of Cold War Research* (Cambridge, MA: Harvard University Press, 2015); Kelly Moore, *Disrupting Science: Social Movements, American Scientists, and the Politics of the Military, 1945–1975* (Princeton, NJ: Princeton University Press, 2013); Wisnioski, *Engineers for Change*, 41–65.

28. Hollomon and Schön would later reunite at MIT, see chapter 5.

29. Robert Gillette, "Scientists in Politics: A Late Entry for Nixon's Group," *Science* 178, no. 4059 (October 27, 1972): 375–77. Engelbert Kirchner and Nina Laserson, "Technology's New Political Environment," *Innovation* (1972): 2–27.

30. "Your Partner in the Hunt for New Products," *Nation's Business*, May 1972, 68–71; Arlen J. Large, "The Big Lawnmower Plot," *New York Times*, April 19, 1974, 8.

31. Richard Hebert and Ralph Hoar, Jr., *Government and Innovation: Experimenting with Change, the Final Report of the Experimental Technology Incentives Program* (Washington, DC: U.S. Department of Commerce, National Bureau of Standards, December 1982). Florence Essers and Jacob Rabinow, eds., *The Public Need and the Role of the Inventor: Proceedings of a Conference Held in Monterey, California, June 11–14, 1973* (Washington, DC: United States Government Printing Office, 1974).

32. Richard Nixon, "Special Message to the Congress on Science and Technology," Online by Gerhard Peters and John T. Woolley, The American Presidency Project https://www.presidency.ucsb.edu/node/255178.

33. Fiona Murray et al., "Grand Innovation Prizes: A Theoretical, Normative, and Empirical Evaluation," *Research Policy* 41 (2012): 1779–1792.

34. Both the British Royal Society and the French Academy of Sciences initiated awards for prominent achievements in science and commerce. See, for example, David Cahan, "The Awarding of the Copley Medal and the 'Discovery' of the Law of Conservation of Energy: Joule, Mayer, and Helmholtz Revisted," *Notes and Records of the Royal Society of London* 66, no. 2 (2012): 125–39.; M. Yakup Bektas and Maurice Crosland, "The Copley Medal: The Establishment of a Reward System in the Royal Society, 1731–1839," *Notes and Records of the Royal Society of London* 46, no. 1 (January 1992): 43–76.

35. In 1955, President Eisenhower proposed an award to honor civilian achievement. But political disagreement about what kind of achievement should be awarded stalled development until the launch of Sputnik made science the most potent symbol of national progress. Robert P. Lowman and Ludy T. Benjamin Jr., "Psychology and the National Medal of Science," *American Psychologist* 67, no. 3 (2012): 174–183.

36. The NSF initially wanted nothing to do with the prize, but when the White House could not find funds for an ambitious $10 million program for multiple winners, it raided $300,000 from the NSF's budget, making the agency a key player in the caper.

37. Memorandum from Lewis Branscomb to James Wakelin, "Prize for Outstanding R&D Accomplishments to Stimulate R&D," March 2, 1972, Office of the Secretary, Executive Secretariat's Subject File 1953–1974, box 278, Sci. and Tech. Assistant Sec., Commerce RG 40.

38. Social innovations were also considered in the planning process, including consumer protection, prison reform, drug rehabilitation, equal opportunity, and welfare programs encouraging "individual initiative." But these "nontechnical" areas were changed to information processing and other technical areas that mirrored the Charpie Report.

39. There was some precedent for defining institutions as innovators. In 1969, the Department of Commerce had proposed a precursor to the prize, a presidential-level award honoring US states or organizations for uses of technology with "the greatest economic impact on mankind." The United Kingdom also had recently initiated an award for corporate innovation. "The President's Award for Outstanding Excellence," March 1969, Assistant Secretary for Science and Technology, box 6, Office of State Technical Services, Commerce RG 40.

40. Remarks by Dr. Edward E. David, Jr., Director, Office of Science and Technology, to the Panel on Presidential Prizes for Innovation, July 24, 1972, Summary of Activities of PPI-72 Staff, box 1, PPI OSTP RG 359.

41. Emphasis in original. Memorandum from Ray Waldmann to John Ehrlichman Re: R&D Prizes, March 6, 1972, box 1, PPI OSTP RG 359.

42. The team included F. G. O'Brien, NSF mathematician; O. M. Bizzell, AEC chemist; W. B. Foster, EPA meteorologist; H. P. Davis, NASA mechanical engineer; C. R. Brewer, HEW; and Carl Muelhause, from Commerce.

43. Memorandum from George Arnstein to Carl Muelhause Re: Pending Problems, PPI Operation, June 5, 1972, Miscellaneous Staff Files, box 1, Correspondence Chron, PPI OSTP RG 359.

44. The Committee was chaired by Ivan L. Bennett, Jr. from New York University Medical Center, and included Costas Anagnostopoulos, Monsanto Company; Lawrence Biebel, a patent attorney; Robert Charpie, president of Cabot Corporation; Lloyd M. Cooke, of Union Carbide; Richard Hall, McCormick and Company; Kenneth G. McKay, AT&T; William B. McLean, US Naval Undersea Warfare Center; David Ragone, Dartmouth; and Gerald F. Tape, President of Associated Universities, Inc.

45. Memorandum from Richard A. Wahl to Daniel V. De Simone Re: Patent Office Review, August 2, 1972, Miscellaneous Staff Files, box 2, Miscellaneous Information re: Patents, PPI OSTP RG 359.

46. The finalists not chosen as winners were Shell Oil's Bruce Collipp and his team for off-shore oil production; Rockefeller Institute's George Cotzias for Parkinson's treatment Levodopa; University of Minnesota inventor E.W. Davis for taconite production; MIT's Charles S. Draper for internal guidance; the Bell Labs team of Carl Frosch, Lincoln Derick, and Jules Andrus for semiconductor fabrication; Edwin Land of Polaroid for instant photography; the MIT Electronic Systems Laboratory for numerical control of machine tools; Corning's Donald Stookey for photosensitive glass; Jerrold Zacharias of the Physical Science Study Committee for his science curricula; and Walter H. Zinn for the boiling water nuclear reactor.

47. Eric S. Hintz, "Portable Power: Inventor Samuel Ruben and the Birth of Duracell," *Technology and Culture* 50, no. 1 (2009): 24–57.

48. Joseph C. O'Mahoney, *The Patent System and the Modern Economy: Study of the Subcommittee on Patents, Trademarks, and Copyrights of the Committee of Judiciary, United States Senate* (Washington, DC: United States Government Printing Office, February 7, 1957).

49. "Presidential Prizes for Innovation 1972: Recommendations to the President by the Advisory Committee," 1972, Nominator Case Files, box 1, PPI OSTP RG 359.

50. Frustrated, Rosen tried to start his own company, but could not raise venture capital. "Presidential Prizes for Innovation 1972."

51. Richard A. Rettig, "Origins of the Medicare Kidney Disease Entitlement: The Social Security Amendments of 1972" in *Biomedical Politics*, ed. Kathi E. Hanna (Washington, DC: National Academy Press, 1991), 176–214.

52. Perry Adkisson and James Tumlinson, *Edward F. Knipling: 1909–2000* (Washington, DC: The National Academies Press, 2003), 8.

53. Marshall Gall, "The Insect Destroyer: Portrait of a Scientist," in *Science for Better Living: The Yearbook of Agriculture 1968* (Washington, DC: United States Department of Agriculture, 1968), 54–57.

54. "Presidential Prizes for Innovation 1972."

55. "Pathfinder," *Think Magazine* (July–August, 1979): 18–24.

56. John W. Backus and Harlan Herrick, "IBM 701 Speedcoding and Other Automatic-Programming Systems," in *Symposium on Automatic Programming for Digital Computers* (Washington, DC: Office of Naval Research, Department of the Navy, 1954), 106–113.

57. Joan Ganz Cooney, "*Sesame Street* Opens Door to Knowledge," *New York Times*, January 12, 1970, 74.

58. Robert W. Morrow, *Sesame Street and the Reform of Children's Television* (Baltimore, MD: Johns Hopkins University Press, 2006), 145.

59. Morrow, *Sesame Street and the Reform of Children's Television*, 77.

60. The CTW touted *Sesame Street*'s scientific basis within a commercial medium. The show sought to instill basic skills like classification, number relations, and reasoning by inference. The CTW, moreover, conducted real-time experiments with a device called the "distractor," which tested children' concentration while viewing pilot segments. It also partnered with the Education Testing Service, the research firm best known for the SAT, to evaluate the show's impact on students. Morrow, *Sesame Street and the Reform of Children's Television*, 158.

61. One educational scholar attempted to create an experiment showing that "poverty children" did no better after being exposed to the show than those receiving in person instruction that emphasized emotional and social development. Herbert A. Sprigle, "Who Wants to Live on Sesame Street?" *Childhood Education* 49, no. 3 (1972): 159–165.

62. Morrow, *Sesame Street and the Reform of Children's Television*, 331.

63. In the wake of Sputnik, Zacharias marshaled a network of scientists, teachers, commercial publishers, and film distributors in support of a "discovery approach" of hands-on investigation. By 1964, over 4,000 teachers and 40 percent of all high school physics students in the country were using materials developed by his Physical Sciences Study Committee. Panel on Educational Research and Development, *Innovation and Experiment in Education* (Washington, DC: 1964).

64. Jack S. Goldstein, *A Different Sort of Time: The Life of Jerrold R. Zacharias, Scientist, Engineer, Educator* (Cambridge, MA: MIT Press, 1992), 300.

65. The committee went out of its way to challenge Cooney's sole selection, making the show the only team winner. Morrisett was instrumental in the show's success, but John Macy, head of the Corporation for Public Broadcasting emphasized Cooney's singular role.

66. D. S., "Whatever Are the Presidential Prizes?" *Science* 178, no. 4067 (December 22, 1972): 1271.

67. Behind the scenes he questioned whether the NSF had the legal authority to award cash prizes, which prompted a terse note from Nixon authorizing the NSF to

do so. Memorandum from Richard Nixon to H. Guyford Stever Re: National Science Foundation Act, April 13, 1972, Entry 31, Correspondence and Subject Files FY73, 78–0008, box 4, White House General File 1972–1973, NSF RG 307.

68. Daniel V. De Simone, *A Metric America: A Decision Whose Time Has Come* (Washington, DC: U.S. Department of Commerce, National Bureau of Standards, 1971).

69. John Walsh, "Science Politics: An Invitation from the White House," *Science* 182, no. 4110 (October 26, 1973): 365–368.

70. The first major international exchange and spread of these ideas was an NAE symposium in 1968 that broad together a small group of leaders that included France's Pierre Aigrain, Japan's Koji Kobayashi, and Texas Instrument's Patrick Haggerty. National Academy of Engineering, *The Process of Technological Innovation* (Washington, DC: National Academy of Sciences, 1969). See also Michael Shanks, *The Innovators: The Economics of Technology* (Baltimore, MD: Penguin, 1967); Organization for Economic Co-Operation and Development, *The Conditions for Success in Technological Innovation* (Paris 1971); Robert Kirk Mueller, *The Innovation Ethic* (American Management Association, 1971).

71. De Simone, for example, became one of the most vocal advocates, and critics of existing failures, for government innovation programs. See, for example, Daniel De Simone, "Innovation and Technology Assessment," *ACS Symposium Series* (Washington DC: American Chemical Society, 1980).

72. Cooney is the possible exception, and her creation remains a polarizing innovation. *Sesame Street*'s recent purchase by the premium cable network HBO has made it a symbol of the privatization of public goods in innovation's name.

CHAPTER 4

1. Participants paid the equivalent of $3,000 in 2020 dollars. Pat McCurdy, "The Innovative Group Innovates with TV," *Chemical Engineering News* 48, no. 7 (1970): 16–17; A. J. Parisi, "New Kind of Conference Focuses on New Ideas in Technology," *Product Engineering*, March 2, 1970, 22–24; "Top Idea Men Trade Ideas," *Business Week*, January 31, 1970, 32–33.

2. "The Innovation Group!" *Wall Street Journal*, April 25, 1969.

3. Fred Turner, *From Counterculture to Cyberculture: Stewart Brand, the Whole Earth Network, and the Rise of Digital Utopianism* (Chicago, IL: University of Chicago Press, 2006); John Markoff, *What the Dormouse Said: How the Sixties Counterculture Shaped the Personal Computer Industry* (New York: Viking, 2005); and Richard Barbrook and Andy Cameron, "The California Ideology," *Science as Culture* 6, no. 6 (1996): 44–72.

4. For a detailed account of the region's rise to power see Margaret O'Mara, *The Code: Silicon Valley and the Remaking of America* (New York: Penguin Press, 2019).

5. Historians of science have utilized the concepts of the *scientific persona* and *scientific self* to explore the co-development of scientists' public representations with scientists' own aspirations of what constitutes a successful career. See Shapin, *The Scientific Life*; Lorraine Daston and H. Otto Sibum, "Introduction: Scientific Personae and Their History," *Science in Context* 16, no. 1–2 (2003): 2; Francesca Bordogna, "Scientific Personae in American Psychology: Three Case Studies," *Studies in the History and Philosophy of Biology and the Biomedical Sciences* 26 (2005): 95–134; Lorraine Daston and Peter Galison, *Objectivity* (Brooklyn, NY: Zone Books, 2007); Paul White, "Darwin's Emotions: The Scientific Self and the Sentiment of Objectivity," *Isis* 100, no. 4 (December 2009): 811–826.

6. See especially, Anne Secord, "'Be What You Would Seem to Be': Samuel Smiles, Thomas Edward, and the Making of a Working-Class Scientific Hero," *Science in Context* 16, no. 1–2 (2003): 147–173.

7. Maurice Holland and Henry F. Pringle, *Industrial Explorers* (New York: Harper and Brothers, 1929).

8. Margaret Mead and Rhoda Métraux, "Image of the Scientist among High-School Students," *Science* 126, no. 3270 (1957): 384–390.

9. Wisnioski, *Engineers for Change*, 16–25.

10. David Kaiser, "The Postwar Suburbanization of American Physics," *American Quarterly* 56, no. 4 (December 2004): 851–888.

11. Steven Shapin, "Who is the Industrial Scientist? Commentary from Academic Sociology and from the Shop-Floor in the United States, ca. 1900–1970," in *The Science-Industry Nexus: History, Policy, Implications*, ed. Karl Grandin, Nina Wormbs, Anders Lundgren, and Sven Widmalm (Sagamore Beach, MA: Science History Publications, 2004), 337–363.

12. Bruce V. Lewenstein, "Magazine Publishing and Popular Science after World War II: How Magazine Publishers Tried to Capitalize on the Public's Interest in Science and Technology," *American Journalism* 6, no. 4 (1989): 218–234.

13. Bruce V. Lewenstein, "The Meaning of 'Public Understanding of Science' in the United States after World War II," *Public Understanding of Science* 1, no. 1 (1992): 45–68.

14. "An Announcement to Our Readers," *Scientific American* 177 (December 1947): 244.

15. William G. Maass, "From the Publisher: A Word of Introduction," *International Science and Technology* 1 (January 1962): front insert.

16. William G. Maass, "New Information Services from a Not-So-Old Publishing House," *Journal of Chemical Documentation* 2, no. 1 (1962): 46–48. William G. Maass, "From the Publisher: After a Year," *International Science and Technology* 12 (December 1962): front insert. This cultivation of exclusivity through "controlled

circulation" was an extension of a practice that Conover Mast first innovated in the late 1920s.

17. Executive editor Daniel Cooper was a nuclear physicist with a PhD from MIT, who left a job at Bell Labs to become managing editor of *Nucleonics*. Among *IST*'s associate editors, David Allison joined after graduating from Rensselaer Polytechnic Institute with a degree in industrial management; Ford Park was an MIT mechanical engineering graduate, former professor at the University of Buffalo, and former editor at *Product Engineering*; Seymour Tilson was a lecturer of geosciences at New York University; and Ted Melnechuk was a chemist and poet who had studied classics with his friend Allen Ginsberg.

18. R. Rodger Remington and Robert S. P. Fripp, *Design and Science: The Life and Work of Will Burtin* (London: Lund Humphries, 2007).

19. *IST* sold limited runs of its cover prints to readers; and, in 1963, the originals were displayed in IBM's corporate art gallery on 57th Street.

20. Daniel I. Cooper, "Pop Science," *American Documentation* (April 1966): 53–56; *The Way of the Scientist* (New York: Simon & Schuster, 1967), 7–8.

21. Donald A. Schön, "Innovation by Invasion," *International Science and Technology* 27 (March 1964): 52–61; Donald A. Schön, "The Fear of Innovation," *International Science and Technology* 27 (March 1966): 70–78.

22. Jack A. Morton, "From Research to Technology," *International Science and Technology* (May 1964): 82–92.

23. Jack A. Morton, "The Microelectronics Dilemma," *International Science and Technology* (July 1966): 35–44

24. William G. Maass, "From the Publisher: Janus' Other Face," *International Science and Technology* 25 (January 1964): front insert.

25. William G. Maass, "From the Publisher: Change or Die!" *International Science and Technology* 35 (November 1964): 9. William G. Maass "From the Publisher: Three Exciting Years," *International Science and Technology* 24 (1963): 6.

26. Barry Commoner, "The Eroding Integrity of Science," *International Science and Technology* 70 (1967): 51–60.

27. Robert Colborn, "In Our Opinion," *International Science and Technology* 70 (1967): 35; and Robert Colborn, "In Our Opinion," *International Science and Technology* 79 (1968): 17.

28. Paul Goodman, "The Case Against Technology," *Innovation* 2 (June 1969): 36–47.

29. For comparison, membership in the American Society of Mechanical Engineers was $20 and a subscription to *Fortune* magazine was $14 in 1969.

30. Robert Colborn, "Says the Editor," *Innovation* 1 (1969): 1.

31. David Allison, "Introduction," in *Dealing with Technological Change: Selected Essays from Innovation, the Magazine about the Art of Managing Advancing Technology* (Auerbach Publishers, 1971). See also David Allison, "Measuring the Good and the Bad of New Technology," *Innovation* 9 (November 1970): 44–55; Michael F. Wolff, "Says the Editor," *Innovation* 17 (January 1971): 1.

32. Donald A. Schön, *Beyond the Stable State* (New York: Random House, 1971), 10–18.

33. See Robert Jay Lifton, "Protean Man," *Archives of General Psychiatry* 24 (1971): 298–304.

34. Schön, *Technology and Change*, 204–216.

35. "11even—Retrieval," *Innovation* 31 (May 1972): 63–64; and Michael F. Wolff, "9ine—Response," *Innovation* 22 (June 1971): 62.

36. Frieda B. Libaw, "And Now, the Creative Corporation," *Innovation* 23 (March 1971): 2–13.

37. Donald Schön, "The Diffusion of Innovation," *Innovation* 5 (October 1969), reprinted in *Managing Advancing Technology: Strategies and Tactics of Product Innovation* (AMA, 1972), 3–20.

38. Nilo Lindgren, "Building a Rational Two-Headed Monster: The Management Style of Robert Noyce and Gordon Moore," *Innovation* (1970). Technology Communication's account became the foundation for future studies of Intel and its founders. See, for example, Leslie Berlin, *The Man Behind the Microchip: Robert Noyce and the Invention of Silicon Valley* (Oxford, UK: Oxford University Press, 2005) which draws extensively on its reportage.

39. Jack A. Morton, *Organizing for Innovation: A Systems Approach to Technical Management* (New York: McGraw-Hill, 1971), 164–165.

40. To achieve this goal, the company also forged a partnership with the American Management Association to reprint its articles in a series of books such as *Managing Advancing Technology. Innovation, Dealing with Technological Change* (Princeton, NJ: Auerbach, 1971); *Innovation, Decision Making in a Changing World* (Princeton, NJ: Auerbach, 1971); and *Innovation, Managing Advancing Technology*, 2 vols. (New York: American Management Association, 1972).

41. The demo can be viewed in full at the Doug Engelbart Institute: https://www.dougengelbart.org/content/view/374/464.

42. Robert Colborn, "11even—Response," *Innovation* 4 (1969): 78–79.

43. 11even—Retrieval," *Innovation* 31 (May 1972).

44. Nilo Lindgren, "The Splintering of the Solid-State Electronics Industry," *Innovation* 8 (1969): 2–15.

45. Lindgren, "The Splintering of the Solid-State Electronics Industry," 51.

46. See, for example, Robert O. Burns, *Innovation: The Management Connection* (Lexington, MA: Lexington Books, 1975) which frequently cited writers in the Innovation Group network.

47. David Silverstein, Philip Samuel, and Neil DeCarlo, *The Innovator's Toolkit: 50+ Techniques for Predictable and Sustainable Organic Growth* (Hoboken, NJ: John Wiley and Sons, 2011); and Steven Johnson, *The Innovator's Cookbook: Essentials for Inventing What's Next* (New York: Riverhead Books, 2011).

48. Gene Bylinsky, *The Innovation Millionaires: How They Succeed* (New York: Charles Scribner's Sons, 1976), xiv.

49. Bylinsky, *The Innovation Millionaires*, 3–24, 73–94.

50. Bylinsky, *The Innovation Millionaires*, 121–142.

51. Hoefler offered a "behind-the-scenes report of the men, money, and litigation that spawned 23 companies—from the fledging rebels of Shockley Transistor to the present day." Don Hoefler, "Silicon Valley U.S.A." *Electronic News* (1971); Turo Uskali and David Nordfors, "The Role of Journalism in Creating the Metaphor of Silicon Valley," https://citeseerx.ist.psu.edu/document?repid=rep1&type=pdf&doi=beaa732aea2d21871b732e9cae468531a270de63.

52. In 1977, local Palo Alto photographer Carolyn Caddes, a student of Ansel Adams, captured a similar sense of community stability in her book *Portraits of Success*. Trading *Innovation's* Day-Glo aesthetic for black and white, the book was a family album of Silicon Valley's most prominent denizens. Caddes even recreated the iconic 1959 group photo of the Fairchild Eight. The company—and region's—pioneers were three decades older with "greying hair, a few bald spots, and several paunches," but otherwise "like a group of overgrown boys." Carolyn Caddes, *Portraits of Success: Impressions of Silicon Valley Pioneers* (Palo Alto, CA: Tioga, 1986), 36–37, 108.

53. See, for example, "California's Silicon Valley is Mass-Producing Millionaires in the Chips," *Life Magazine*, March 1982, 32–38.

54. S. H. Price, "To Newsweek/Access Readers," *Newsweek Access*, Fall 1984, 4.

55. David Sheff, "Steven Jobs," *Playboy*, February 1, 1985: 49–54, 58, 70, 174–184.

56. Tom Zito, "Steve Jobs Explains," *Newsweek Access*, Fall 1984, 42–51.

CHAPTER 5

1. Randy Rieland, "Experiment Refuels Cab Firm," *Pittsburgh Press*, December 1, 1975, 1; Tung Au and Dwight M. B. Baumann, *A Multi-Origin/Destination Taxi System: Preliminary Design for Single-Origin/Destination Trips* (Washington, DC: Department of Transportation, 1975).

2. Mary Ann Scheirer et al., *Innovation and Enterprise: A Study of NSF's Innovation Centers Program* (Rockville, MD: Westat, Inc., 1985).

3. The literature on university innovation, technology transfer, and entrepreneurship education is vast. See, for example, Gary D. Libecap, *University Entrepreneurship and Technology Transfer: Process, Design, and Intellectual Property* (London: Elsevier, 2005); Albert N. Link, Donald S. Siegel, and Mike Wright, eds., *The Chicago Handbook of University Technology Transfer and Academic Entrepreneurship* (Chicago, IL: University of Chicago Press, 2014).

4. Donald G. Stein, ed., *Buying in or Selling Out* (New Brunswick, NJ: Rutgers University Press, 2004); Philip Mirowski, *Science-Mart: Privatizing American Science* (Cambridge, MA: Harvard University Press, 2011); Christopher Newfield, *Ivy and Industry: Business and the Making of the American University, 1880–1980* (Durham, NC: Duke University Press, 2004).

5. Popp Berman, *Creating the Market University*.

6. Daniel Kleinman, *Politics on the Endless Frontier* (Durham, NC: Duke University Press, 1995).

7. Cyrus Mody shows how these two dimensions combined at Stanford where technological experiments with new MOSFET transistors coincided with institutional experiments designed to counter local political and economic crises. Cyrus Mody, *The Long Arm of Moore's Law: Microelectronics and American Science* (Cambridge, MA: MIT Press, 2017). For more on the material changes, see Christophe Lecuyer, *Making Silicon Valley: Innovation and the Growth of High Tech, 1930–1970* (Cambridge, MA: MIT Press, 2005).

8. US Department of Commerce, *Technological Innovation*.

9. David F. Noble, *America by Design: Science, Technology, and the Rise of Corporate Capitalism* (New York: Alfred A. Knopf, 1977).

10. Margaret B.W. Graham, "Entrepreneurship in the United States, 1920–2000," in *The Invention of Enterprise: Entrepreneurship from Ancient Mesopotamia to Modern Times*, ed. David S. Landes, Joel Mokyr, and William J. Baumol (Princeton, NJ: Princeton University Press, 2010), 401–442; Jerome A. Katz, "The Chronology and Intellectual Trajectory of American Entrepreneurship Education 1876–1999," *Journal of Business Venturing* 18 (2003): 283–300.

11. James F. Mayar and Dean C. Coddington, "Academic Spin offs," *Industrial Research* 7 (April 1965): 62–71; Arnold C. Cooper, "Incubator Organizations, Spin-Offs, and Technical Entrepreneurship," *Indiana Academy of the Social Sciences Proceedings* 4 (April 1970): 26–33.

12. James W. Schreier and John L. Komives, *The Entrepreneur and New Enterprise Formation: A Resource Guide* (Milwaukee, WI: Center for Venture Formation, 1973).

13. Daniel V. De Simone, *Education for Innovation* (London: Pergamon Press, 1968), 6, 28–30, 90–92, 147–149.

14. Dian Olson Belanger, *Enabling American Innovation: Engineering and the National Science Foundation* (West Lafayette, IN: Purdue University Press, 1998); Leslie, *The Cold War and American Science*; Etzkowitz, *The Triple Helix*; Rebecca Lowen, *Creating*

the Cold War University: The Transformation of Stanford (Berkeley, CA: University of California Press, 1997).

15. Jack Renirie, "NSF Establishes R&D Incentives, Assessment Offices; Smith and Lederman Appointed Director," *National Science Foundation: News* (August 16, 1972).

16. Deborah Shapley, *The National Science Board: A History in Highlights, 1950–2000* (Arlington: National Science Board, 2000); Deborah Shapley, "Technology Incentives: NSF Gropes for Relevance," *Science* 179, no. 4078 (March 16, 1973): 1105–1107; Popp Berman, *Creating the Market University*, 124–126.

17. "Response to OMB Analysis of Experimental R&D Incentives," RG 307, Records of the Office of the National Science Foundation, Dep. Dir-Experiment R&D Inc. Files, box 1, Experimental Incentives, National Archives at College Park, Maryland (hereafter cited as NSF RG 307).

18. From the start, the experimental nature of the innovation centers was called into question. A government report from the Office of Budget and Management, for example, argued that it could not be "run as a true experiment under control conditions." "Analysis of NSF Incentives Plan," June 13, 1973, Dep. Dir-Experiment R&D Inc. Files, box 1, Experimental Incentives, NSF RG 307. Gerald G. Udell, Ken Baker, and Robert M. Colton, "The Innovation Center Program," *Research Management* 22, no. 4 (July 1979): 32–38.

19. Y.T. Li, "General Statement on the Operation Plan of the MIT Innovation Program," September 12, 1973, AC0012, box 58, folder Innovation Program 1 of 3. Department of Distinctive Collections, MIT Libraries, Cambridge, Massachusetts.

20. David Kaiser, ed., *Becoming MIT: Moments of Decision* (Cambridge, MA: MIT Press, 2010); Wisnioski, *Engineers for Change*, 178–183; Benson R. Snyder, *The Hidden Curriculum*, (Cambridge, MA: MIT Press, 1973).

21. Yao Tzu Li, *Freedom and Enlightenment: My Life as an Educator/inventor in China and the United States* (Lexington, MA: The Lexington Press, 2003).

22. Yao Tzu Li, "A Proposal of a Supplementary Education System for the Training of Innovators and Entrepreneurs," December 1972, 11, AC0012, box 58, folder: Innovation Program 2 of 3, Department of Distinctive Collections, MIT Libraries, Cambridge, Massachusetts.

23. Li, "A Proposal of a Supplementary Education System for the Training of Innovators and Entrepreneurs," 11–25.

24. "Incubators for Entrepreneurs," *Mosaic* (July/August 1978), 14.

25. "MIT Innovation Program Operating Plan," December 1973, 19, AC0012, box 58, folder, Innovation Program 1 of 3, Department of Distinctive Collections, MIT Libraries, Cambridge, Massachusetts.

26. "MIT Innovation Program Operating Plan," 19; Yao Tzu Li, "The Structure and Operation of the MIT Innovation Center: Past Experience and Future Prospects," in *Innovation and Innovation Centers: 1978 Proceedings of the Symposium on Innovation*

and Innovation Centers Held at Cambridge, MA on May 17–19, 1978 (Washington, DC: National Science Foundation, 1978), 129–135.

27. "Incubators for Entrepreneurs," 11, 14. Philip Doucet, "Developing a New Company Under the Innovation Co-Op Program," in *Innovation and Innovation Centers,* 156–158.

28. "MIT Development Foundation, Inc." AC 118, Office of the President and Chairman of the Corporation (Johnson), box 209 MIT Development Foundation, Department of Distinctive Collections, MIT Libraries, Cambridge, Massachusetts. One of Morison's earliest mentees included Robert A. Swanson, a chemistry undergraduate who petitioned to take courses at the Sloan School and who went on to cofound Genentech. Sally Smith Hughes, *Genentech: The Beginnings of Biotech* (Chicago, IL: University of Chicago Press, 30).

29. Irwin Feerst, "University Innovation Centers: Burnout or Bright Idea?" *New Engineer* (March 1978), 14–28; Li, "The Structure and Operation of the MIT Innovation Center," 133.

30. Scheirer, et al., *Innovation and Enterprise,* 11–17. For details on the latter years of the Center and conflicts between Hollomon and Li, see correspondence in MC0355 J. Herbert Hollomon Papers, box 2, folders MIT Innovation Center: General, 1974–1981 and MIT Innovation Center, 1978–1979, Department of Distinctive Collections, MIT Libraries, Cambridge, Massachusetts.

31. Udell, Baker, and Colton, "The Innovation Center Program," 33.

32. Essers and Rabinow, eds., *The Public Need and the Role of the Inventor.* For a recent version of the argument see Jack Hitt, *Bunch of Amateurs: A Search for the American Character* (New York: Crown, 2012).

33. This aspect of the program followed on the heels of the Department of Commerce's State Technical Services program described in chapter 3.

34. Gerald G. Udell, Kenneth G. Baker, and Gerald S. Albaum, "Creativity: Necessary, but Not Sufficient," *The Journal of Creative Behavior* 10, no. 2 (1976): 92–103.

35. Scheirer et al., *Innovation and Enterprise,* 29–30.

36. Gerald G. Udell, Michael F. O'Neill, and Kenneth G. Baker, *Guide to Invention and Innovation Evaluation* (Washington, DC: National Science Foundation, RANN-Research Applied to National Needs, Division of Intergovernmental Science and Public Technology, 1977); Kenneth G. Baker, "The Center for the Advancement of Invention and Innovation University of Oregon," in *Innovation and Innovation Centers,* 115–128.

37. Douglas Colligan, "Help for Inventors: Innovation Center Gives Cheap, Expert Advice," *Popular Science* 212, no. 5 (1978): 92–94, 182.

38. Udell, Baker, and Colton, "The Innovation Center Program," 35.

39. Paul Rolly, "Innovation Mandate," *Venture: The Magazine for Entrepreneurs* (1979): 32–35.

40. WIN later became the World Innovation Network to dispel assumptions that it was part of Wal-Mart. Nancy Bowman-Upton, Samuel L. Seaman, and Donald L. Sexton, "Innovation Evaluation Programs: Do They Help the Inventors?" *Journal of Small Business Management* 27, no. 3 (July 1989): 23–30.

41. Edwin T. Layton, "American Ideologies of Science and Engineering," *Technology and Culture* 17, no.4 (1976): 688–701.

42. "Incubators for Entrepreneurs," 12.

43. "CED Readiness Assessment Tips," Center for Entrepreneurial Development Brochure. box 8, folder 30, Carnegie Institute of Technology, Dean's Subject Files, Carnegie Mellon University Archives (hereafter cited as CMU Dean's Files).

44. The system utilized existing electrical wiring to create an easy to install and responsive security environment. Feerst, "University Innovation Centers," 15–16.

45. Lea M. Griswold, "Compuguard Corporation: The First Five Years," in *Innovation and Innovation Centers*, 173–179; John Mortimer and Brian Rooks, *The International Robot Industry Report* (Berlin, Germany: Springer-Verlag, 1987), 18–21; Mrinalini Krishna, "Q&A With Immigrant Billionaire Romesh Wadhwani: America Needs Immigrants," *Forbes*, October 5, 2016, https://www.forbes.com/sites/mrinalinikrishna/2016/10/05/qa-with-immigrant-billionaire-romesh-wadhwani-america-needs-immigrants-2/?sh=52dc6bb51b2b.

46. Herb Toor to Bob Kaplan, "CED Notes," March 26, 1979; Dwight Baumann to Richard Cyert and Deans, "Wharton School—Management and Technology Programs," June 9, 1980, W.A. Sirignano to Dwight Baumann, February 2, 1981, boxes 8 and 11, CMU Dean's Files.

47. U.S. Congress, House of Representatives, Committee on Science and Technology, Subcommittee on Science, Research, and Technology, *Government and Innovation: University-Industry Relations*, 96th Cong., 1st sess., July 31; August 1–2, 1979, 437. The argument that Salt Lake City was an entrepreneurial venture was first made in Jonathan Hughes, *The Vital Few: The Entrepreneur and American Economic Progress*, 2nd ed. (New York: Oxford University Press, [1965] 1986), 67–120.

48. Jacob Gaboury, "Image Objects: An Archeology of 3D Computer Graphics, 1965–1979," (PhD Diss. New York University, 2014).

49. Warren S. Brown, "The Innovation Center University of Utah," in *Innovation and Innovation Centers*, 192–200.

50. A few years after the Center started, Jacobsen founded the robotics company Sarcos, later Raytheon-Sarcos, that would later create the mechanized dinosaurs for the *Jurassic Park* ride at Universal Orlando, animatronic pirates for Disney's *Pirates of the Caribbean* rides, and the controllers for the Bellagio fountains in Las Vegas. "In Memoriam: Stephen C. Jacobsen," *The University of Utah Engineering News*, University of Utah, April 20, 2016, https://www.coe.utah.edu/jacobsen.

51. Clayton Jones, "University Tries to Incubate a Few Good Inventions," *Christian Science Monitor*, April 28, 1982, https://www.csmonitor.com/1982/0428/042864.html.

52. Wayne S. Brown and William R. Bowen, "Editorial," *Technovation* 1, no. 1 (1981): 1.

53. Scheirer, et al., *Innovation and Enterprise*, 36–38.

54. Rod Willis, "What Should Be the Federal Role in Startups?" *Management Review* 74, no. 11 (1985): 11–13.

55. Robert M. Burger, *An Analysis of the National Science Foundation's Innovation Centers Experiment: An Effort to Promote Technological Innovation and Entrepreneurship in Academic Settings* (Washington, DC: National Science Foundation, 1977), 26.

56. Rolly, "Innovation Mandate," 35.

57. Robert M. Colton, "Existing Centers: Their History, Future and Roles of the University," in *Innovation, Entrepreneurship and the University: Proceedings of a Conference held at University of California Santa Cruz, CA, November 8-9-10, 1978*, ed. Narinder S. Kapany (Santa Cruz: Center for Innovation and Entrepreneurial Development, 1978), 10–11.

58. "Invitation to Center for Entrepreneurial Development Reception," April 14, 1980, box 8, folder 30, CMU Dean's Files.

59. Press Release from the Office of U.S. Senator William Proxmire, March 14, 1975, AC008, box 199, folder Office of the President, Department of Distinctive Collections, MIT Libraries, Cambridge, Massachusetts.

60. Feerst, "University Innovation Centers," 27–28.

61. Julian Camacho and Grant Venerable, "Nontraditional Students," in *Innovation, Entrepreneurship and the University*, 114.

62. Martin D. Robbins and J. Gordon Milliken, "Government Policies for Technological Innovation: Criteria for an Experimental Approach," *Research Policy* 6 (1977): 227.

63. R. M. Colton, "Rejoinder to Government Policies for Technological Innovation: By Robbins and Milliken," *Research Policy* 6 (1977): 249.

64. Martin D. Robbins and J. Gordon Milliken, "Government Policies for Technological Innovation: Criteria for an Experimental Approach," *Research Policy* 6 (1977): 227.

65. Scheirer, et al., *Innovation and Enterprise*, 1.

66. These Centers were to enhance innovation by (1) facilitating university–industry cooperation, (2) developing a "generic research base" in areas in which "individual firms have little incentive to invest," (3) training students in the process of technological innovation, and (4) improving mechanisms for disseminating technical information among universities and industry. The act also required all federal laboratories to establish an office of technology transfer and established a National Technology Medal. Popp Berman, *Creating the Market University*, 133.

67. Subcommittee on Science, Research, and Technology of the Committee on Science and Technology, U.S. House of Representatives, *Summary of House and Senate Hearings on Government-University-Industry Relations* (Washington, DC: US Government Printing Office, 1980), 42–43.

68. Scheirer, et al., "Innovation and Enterprise," xxii, 41–71.

69. "Incubators for Entrepreneurs," 11.

70. Errol Arkilic, "Raising the NSF Innovation Corps," in Wisnioski, Hintz, and Kleine, *Does America Need More Innovators?* 73–75.

71. Dwight M. Baumann, "Center for Entrepreneurial Development, Carnegie-Mellon University," in *Innovation, Entrepreneurship and the University*, 19.

72. Thomas H. Byers, Richard C. Dorf, and Andrew J. Nelson, *Technology Ventures: From Idea to Enterprise*, 3rd ed. (New York: McGraw-Hill, 2011), 13.

73. Timothy Sands, "A Message to the Virginia Tech Community from President Timothy Sands," *Virginia Tech News*, November 13, 2015, https://vtnews.vt.edu /articles/2015/11/111315-president-presidentletter.html.

CHAPTER 6

1. Michael L. Dertouzos, Robert M. Solow, and Richard K. Lester, *Made in America: Regaining the Productive Edge* (Cambridge, MA: MIT Press, 1989). Hobart Rowen, "Finding a Way to Thrive Again in a World Economy," *Washington Post*, May 14, 1989, H1.

2. Stewart Brand, *The Media Lab: Inventing the Future at MIT* (New York: Viking Press, 1987).

3. Sebastian Pfotenhauer and Joakim Juhl, "Challenging the "deficit model" of innovation: Framing policy issues under the innovation imperative," *Research Policy* 48, no. 4 (2018): 895–904.

4. For surveys of innovation policy from the 1970s to the 1990s, see Ann Johnson, "The End of Pure Science: Science Policy from Bayh-Dole to the NNI," in *Discovering the Nanoscale*, ed. Davis Baird, Alfred Nordmann, and Joachim Schummer (Amsterdam: IOS Press, 2004), 217–230; Alan I. Marcus and Amy Sue Bix, *The Future is Now: Science and Technology Policy in America Since 1950* (New York: Humanity Books, 2007); Cyrus Mody, *The Squares: US Physical and Engineering Scientists in the Long 1970s* (Cambridge, MA: MIT Press, 2022); Lee Vinsel, "How to give up the I-word, pt. 1," *Culture Digitally*, September 22, 2014, https://culturedigitally.org/2014/09/how -to-give-up-the-i-word-pt-1/; Julia Marino, "Fighting the Cold War and the Market War' through Critical Technologies, 1979–1992," *Historical Studies in the Natural Sciences* 52, no. 4 (2022): 485–523; Godin, *The Idea of Technological Innovation*, 96–114.

5. Simon Ramo, "Memo to Members of the Science and Technology Advisory Groups," November 17, 1975. Associate Director for Energy and Science Files,

1974–77, box 35, Science and Technology Policy, 1975: Advisory Groups 1, Gerald R. Ford Library; Walter Sullivan, "Loss of Innovation in Technology is Debated," *New York Times*, Nov 25, 1976, 44.

6. Center for Policy Alternatives, *National Support for Science and Technology: An Examination of Foreign Experience, Working Paper 75–12* (Cambridge, MA: Center for Policy Alternatives, 1975), 68. In follow-up studies, the CPA pushed for federal investment in civilian technology and government coordination of industry, pointing to the Netherlands and Sweden as successful innovation nations. Meanwhile, advocates for deregulation emphasized the role of private industry in industrial innovation. Shing K. Fung et al., *Government Action and the Innovation Process, CPA-77-3* (Cambridge, MA: MIT Center for Policy Alternatives, 1977).

7. *1980 National Science Foundation Authorization: Hearings on H.R. 2276, First Session, Before the Subcommittee on Science, Research and Technology of the Committee on Science and Technology*, 96th Cong. 603 (1979) (statement of Phil Handler, President, National Academy of Sciences).

8. See, for example, "Vanishing Innovation," *Business Week*, July 3, 1978, 46; "The Sad State of Innovation," *Time*, October 22, 1979, 70–71. Democrat Adlai Stevenson III and republican Strom Thurmond, for example, added "Vanishing Innovation," the latter while using Thomas Edison's 100th birthday as a call for "America's need to revitalize the spirit of American innovation."

9. "Innovation: Has America Lost Its Edge?" *Newsweek*, June 4, 1979, 58–66.

10. Department of Commerce, *Advisory Committee on Industrial Innovation: Final Report* (Washington, DC: Department of Commerce, 1979), iii, 30.

11. On Atari democrats see O'Mara, *The Code*, 192–198. On one of the first, and celebratory, analyses of neoliberals, see Randall Rothenberg, *The Neoliberals: Creating the New American Politics* (New York: Simon & Schuster, 1984).

12. "What the NSF must realize," Penn State president Eric Walker wrote in support of the bill, "is that engineering is not just research; it is innovation." From Eric Walker to Ms. Vernice Anderson, December 2, 1980, box 175, National Science Board 2, Gerald R. Ford Library.

13. Belanger, *Enabling American Innovation*, 139–154.

14. E. E. David, "The Role of Science and Engineering in the Energy Future," November 19, 1981, box 174, National Commission on Excellence in Education—Hearing on Education in Science, Mathematics and Technology, 1982, Gerald R. Ford Library.

15. President's Commission on Industrial Competitiveness, *Global Competition: The New Reality*. Washington, DC: President's Commission on Industrial Competitiveness, 1985).

16. David Ricardo, *On the Principles of Political Economy and Taxation* (London: John Murray, 1817).

17. *Global Competition*, 7, 17–18.

18. *Global Competition*, 31.

19. George N. Hatsopoulos, Paul R. Krugman, and Lawrence H. Summers. "US Competitiveness: Beyond the Trade Deficit." *Science* 241, no. 4863 (1988): 299–307; Philip H. Abelson, "Competitiveness: A Long-Enduring Problem," *Science* 240, no. 4854 (1988): 865. Krugman later swatted away fears of the United States going out of business like a stagnant company, but also pointing to real challenges in key technology sectors. Paul A. Krugman, "Myths and Realities of U.S. Competitiveness," *Science* 254, no. 5,033 (1991): 811–815.

20. Shigeo Minabe, "Japanese Competitiveness and Japanese Management," *Science* 233, no. 4,761 (1986): 301–304.

21. Richard Cyert, David C Mowery, and the Committee on Science, Engineering, and Public Policy US Panel on Technology and Employment, *Technology and Employment: Innovation and Growth in the US Economy* (Washington, DC: National Academy Press, 1987), viii, 4. A similar perspective can be found in Ralph Landau and Nathan Rosenberg, *The Positive Sum Strategy: Harnessing Technology for Economic Growth* (Washington, DC: National Academy Press: 1986).

22. "Science, Technology, and Innovation fact sheet," Bush Quayle, November 3, 1988, box 130, Post Government Subject File Carnegie Commission on Science, Technology, and Government, Gerald R. Ford Library.

23. Richard R. Nelson, "National Innovation Systems: A Retrospective Study," *Industrial and Corporate Change* 1, no. 2 (1992): 347–374.

24. *Made in America*, 3, 35–40.

25. Fred Block, "Innovation and the Invisible Hand of Government," in *State of Innovation: The US Government's Role in Technology Development*, ed. Fred Block and Matthew R. Keller (Boulder: Paradigm, 2011), 10–12; Linda Weiss, *America Inc? Innovation and Enterprise in the National Security State* (Ithaca, NY: Cornell University Press, 2014), 43, 59–64.

26. Mazzucato, *Entrepreneurial State*, 85–92.

27. Grant Black, *The Geography of Small Firm Innovation* (Boston, MA: Kluwer, 2004), 24–25. SBIR has generated significant debate about the ethics and effectiveness of the federal government as a venture capitalist. Champions describe it as the program that solved the competitiveness crisis. See especially, David B. Audretsch "Standing on the Shoulders of Midgets: The U.S. Small Business Innovation Research Program (SBIR)," *Small Business Economics* 20 (2003): 129–135. Critics such as J. David Roessner in contrast have argued that oversight and evaluation of the program were never taken seriously because supporters had too much at stake to question it. J. David Roessner, "Evaluating Government Innovation Programs: Lessons from the U.S. Experience," *Research Policy* 18 (1989): 343–359. For a quantitative analysis of SBIR

effectiveness, see Josh Lerner, "The Government as Venture Capitalist: The Long-Run Impact of the SBIR Program," *The Journal of Private Equity* 3, no. 2 (2000): 55–78.

28. The policy implication was that the tepid results of previous federal jobs programs were due to their capture by large corporations. Yet positive next steps were not obvious because the energy that gave small companies their vitality also made them less than ideal partners. Put more bluntly, Birch hinted that government oversight and coordination of thousands of small companies was near impossible. David Birch, *The Job Generation Process* (Cambridge, MA: MIT Program on Neighborhood and Regional Change, 1979).

29. The U.S. Department of Commerce, for example, created new State and Local Initiatives on Productivity, Technology, and Innovation which documented programs in twenty-six states between 1982 and 1987. Advisory Commission on Intergovernmental Relations, *State and Local Initiatives on Productivity, Technology, and Innovation: Enhancing a National Resource for International Competitiveness* (Washington, DC: Government Printing Office, 1990).

30. "Request for Proposals to Fund Projects through the Ben Franklin Challenge Grant Program," box 2, folder 10, Tepper School of Business Records, Carnegie Mellon University Archives.

31. Innovation experts' stock inside academia rose as they described global-historical changes that conveniently put their own work at the center. Daniel Bell's classic *Coming of Post-Industrial Society* was updated and refined on multiple fronts. A particularly influential frame for this knowledge economy came from Europe where a team of philosophers, management theorists, and science policymakers asserted a shift from traditional "Mode 1" scientific research, detached from societal context, to "Mode 2" research, characterized by *transdisciplinary* projects in society's service. Daniel Bell, *The Coming of Postindustrial Society: A Venture in Social Forecasting* (New York: Basic Books, 1973); Gibbons, et al., *The New Production of Knowledge*.

32. National Science Board, *University-Industry Research Partnerships: Myths, Realities and Potentials, Fourteenth Annual Report of the National Science Board* (Washington, DC: National Science Foundation, 1982), 16.

33. A retrospective study claims that that the ERC program produced nearly 11,000 journal articles, 908 inventions, and 391 patents between 1985 and 2002. Sara Perry, Steven Currall, T. E. Stuart, "The Pipeline from University Laboratory to New Commercial Product: An Organizational Framework Regarding Technology Commercialization in Multidisciplinary Research Centers," in *The Creative Enterprise: Managing Innovative Organizations and People*, Vol. 1, ed. R. D. Shelton et al. (Westport, CT: Preager, 2007), 85–106.

34. On Boyer, see Frederic Golden, "Shaping Life in the Lab," *Time*, March 9, 1981, 36–49; Robert F. Johnson and Christopher G. Edwards, *Entrepreneurial Science: New Links Between Corporations, Universities, and Government* (New York: Quorum Books, 1987).

35. Maryann P. Feldman and Richard Florida, "The Geographic Sources of Innovation: Technological Infrastructure and Product Innovation in the United States," *Annals of the Association of American Geographers* 84, no. 2 (1994): 210–229.

36. Where the technology–economics experts echoed Ricardo, the regional innovation experts looked to another nineteenth-century economist, Alfred Marshall, who had studied the clustering of British factories. Alfred Marshall, *Principles of Economics* (London: Macmillan, 1890).

37. Saxenian, *Regional Advantage*, 7–9.

38. Saxenian, *Regional Advantage*, 167.

39. The closest civic connection was Nixon's space age genius, Harold Rosen, who finally received his award for telecommunications satellites (see chapter 3). These selections were largely about correcting public images of liberal West Coast entrepreneurs; indeed, the only awardees widely known to the public were Apple Computer founders Steve Jobs and Steve Wozniak. In 2007, as part of the America COMPETES Act, George W. Bush, renamed the award the National Medal for Technology and Innovation. "National Medal of Technology and Innovation (NMTI)," https://www.uspto.gov/learning-and-resources/ip-programs-and-awards/national-medal-technology-and-innovation-nmti.

40. Belanger, *Enabling American Innovation*, 211–234.

41. *Made in America*, 21, 131, 157.

42. Matthew Wisnioski, "Why MIT Institutionalized the Avant-Garde: Negotiating Aesthetic Virtue in the Postwar Defense Institute," *Configurations* 21, no. 1 (2013): 85–116; Matthew Wisnioski, "Centerbeam: Art of the Environment," in *A Second Modernism: MIT, Architecture and the 'Techno-Social' Moment*, ed. Arindam Dutta (Cambridge, MA: MIT Press, 2013), 189–225.

43. Daniel Cardoso Llach, *Builders of the Dream: Software and the Imagination of Design*, (London: Routledge, 2015).

44. Brand, *The Media Lab*, 251.

45. Brand, *The Media* Lab, 6, 45, 85, 108.

46. Barry M. Katz. *Make it New: The History of Silicon Valley Design* (Cambridge, MA: MIT Press, 2015), 117–148.

47. Stanford Design Division Brochure, box 1, Stanford Design EXPErience Records, SC0644, Stanford University Special Collections and University Archives (hereafter cited as EXPErience SC0644).

48. James Adams, "Poorly premeditated pre-meetathon thoughts from Adams," October 24, 1969, box 16, Chronological File Sept 1969—August 1970, James L. Adams Papers, SC0949, Stanford University Special Collections and University Archives (hereafter cited as Adams Papers SC0949).

49. William Massy, Memo, "Stanford NSF R&D Incentives," April 10, 1972, box 22, NSF Technology Incentives, Provost's Office, Records, SC0115; Jim Adams, "How I visited Kansas City," box 16, Chronological File Sept 1972—August 1973, Adams Papers SC0949.

50. James Adams, *Conceptual Blockbusting: A Pleasurable Guide to Better Problem Solving* (New York: Norton, 1974); Robert H. McKim, *Experiences in Visual Thinking* (Belmont, CA: Brooks/Cole Publishing, 1972).

51. Stanford Engineering Executive Program, July 1–13, 1979, box 16, Chronological File 1979, Adams Papers SC0949.

52. Kathleen Brandt and Brian Lonsway, "Beanbags and Microscopes at Xerox Park," in *Laboratory Lifestyles: The Construction of Scientific Fictions,* ed. Sandra Kaji-O'Grady, Chris L. Smith, and Russell Hughes (Cambridge, MA: MIT Press 2018); 29–48; Katz, *Make it New,* 55–116.

53. As Kelley tells it, he learned about the program through a carpool ride during the 1973 oil crisis. David M. Kelley, An Oral History Interview conducted by Judee Humburg, Stanford Oral History Program, Stanford Historical Society, 2012.

54. David Kelley to Members of the Stanford "Design Experience" Committee, June 25, 1993, EXPE Final Report, box 1, EXPErience SC0644.

55. Stanford Design Experience 2000, box 1, EXPErience SC0644.

56. "Hasso Plattner Institute of Design 2005–2006," box 3, Institute of Design, Terry Allen Winograd Papers, SC1165, Stanford University Special Collections and University Archives.

57. NSF Grant #GI-29728 June 1973 Second Annual Progress Report of the Processing Research Institute, PRI Processing Research Institute 1973, box 20, folder 15, CMU Dean's Files.

58. CMU PRI Final Report NSF 1977, box 20 folder 15, CMU Dean's Files.

59. "Making a B-school More Relevant," *Business Week,* December 5, 1970, 58–60.

60. The clearest expression of this vision is Herbert Simon, *Sciences of the Artificial* (Cambridge, MA: MIT Press, 1969).

61. A proposal to the Alfred P. Sloan Foundation, June 6, 1979, box 9, folder 44 CMU Dean's Files.

62. Cost Estimate to move CIT to a position of National Leadership," box 9 folder 39, CMU Dean's Files; Alex Poinsett, "School for Urban Troubleshooters," 32, no. 2 *Ebony* (1976): 92–98.

63. Sarah Talukdar, "The Engineering Design Research Center: Some Lessons from its History" *Power Engineering Society Summer Meeting Conference Proceedings* vol. 2 (2001): 1100–1102.

64. *Scientific American* heralded EDRC's 3D-printing research as the coming of "desktop artisans." Gary Stix, "Desktop Artisans," *Scientific American,* 266, no. 4 (1992):

141–142; Susan Finger, Suresh Konda, and Eswaran Subrahmanian, "Concurrent Design Happens at the Interfaces," *AI EDAM* 9, no. 2 (1995): 89–99.

65. CMU's arts-based design also continued in EDRC's shadow but likewise embraced computation. Industrial design professor Joe Ballay defended this craft mentality as relevant for "the electronic, postindustrial era" and saw artist–engineer collaboration as a "test bed for the real, social, human effects of the new technology." J.M. Ballay, "A strategic focus for the department of design," 1981, box 1, Design Annual Report 1980–1981, College of Fine Arts Records, Carnegie Mellon University Archives.

66. The initiative started with support of a $1 million grant from the local Richard King Mellon Foundation and was developed in collaboration with the University of Pittsburgh, the Pittsburgh High Technology Council, Small Business Development Centers, US Small Business Administration, and BFTC.

67. "Enterprise Makes Advances in Helping to Create Much-Needed Jobs in Region," *Enterprise* 9, no. 4 (1992): 3.

68. "Donald H. Jones Center for Entrepreneurship," GSIA Brochure, box 5, folder 21, Tepper School of Business Records, Carnegie Mellon University Archives.

69. Among the additional innovation expertise units was the history and social science-oriented Program on Innovation, Technology and Economic Growth (PITEG).

70. Florida, *Rise of the Creative Class*, 215.

71. Robert Burchell, James Carr, Richard Florida, and James Nemeth, *The New Reality of Municipal Finance: The Rise and Fall of the Governmental City* (New Brunswick, NJ: Rutgers University Press, 1984).

72. Richard L. Florida, "Banking on Housing: The Political Economy of Financial Deregulation and the Reorganization of Housing Finance," (PhD diss. Columbia University, 1986).

73. Andrew Mair, Richard Florida, and Martin Kenney, "The New Geography of Automobile Production: Japanese Transplants in North America," *Economic Geography* 64, no. 4 (1988): 352–373

74. Richard Florida and Martin Kenney, "Venture Capital and High Technology Entrepreneurship," *Journal of Business Venturing* 3 (1988), 316–317.

75. Richard Florida and Martin Kenney, *The Breakthrough Illusion: Corporate America's Failure to Move from Innovation to Mass Production* (New York: Basic Books, 1990), 80–95, 193–194.

76. Regional Economic Revitalization Initiative, *The Greater Pittsburgh Region: Working Together to Compete Globally* (Pittsburgh, PA: Regional Economic Revitalization Initiative, 1994).

77. Lewis Branscomb, Fumio Kodama, and Richard Florida, eds. *Industrializing Knowledge: University-Industry Links in Japan and the United States* (Cambridge, MA: MIT Press, 1999).

78. Florida, *Rise of the Creative Class*, 216.

79. Florida did not use the term *innovators* to describe the creative class but the occupational and personal characteristics were synonymous.

80. Stewart Brand had used this data to explain the goals of the MIT Media Lab. Brand, *The Media Lab*, 8. His data came from Marc Porat, *The Information Economy: Definition and Measurement* (Washington, DC: US Department of Commerce, 1977).

81. Florida, *The Rise of the Creative Class*, 73.

82. Florida, *Rise of the Creative Class*, 45–48.

83. Florida, *Rise of the Creative Class*, 37.

84. Florida, *Rise of the Creative Class*, 71

85. Florida, *Rise of the Creative Class*, xxx, 317.

86. Articles focusing on the Creativity Index numbered in the hundreds. Thank you to Sarvnaz Lotfi for collecting them. Mark Lisheron and Bill Bishop, "City of Ideas: Prosperity and Its Price: Why the Creative Come Here," *Austin American-Statesman*, May 12, 2002; Mark Watson, "Memphis Can Become 'Magnet,' Expert Says," *The Commercial* Appeal, February 23, 2002, C1; Kevin Collison, "Amenities Help, But Open Minds Will Help More," *The Kansas City Star*, April 2, 2002, D20; Froma Harrop, "At the Rate Mountaintop Removal Is Going, West Virginia Will Look More Like Indiana Every Day," *Charleston Gazette*, May 12, 2002, 1C.

87. Edward L. Glaeser, "Review of Richard Florida's *The Rise of the Creative Class*," Regional Science and Urban Economics 35 (2005): 593–596; Steven Malanga, "The Curse of the Creative Class," *City Journal* 14, no. 1 (2004): 36–45; Tom Craig and Maury Walker, "Letters," *Sacramento Bee*, June 8, 2002, B7; Jamie Peck, "Struggling with the Creative Class," *International Journal of Urban and Regional Research* 29, no. 4 (2005): 740–770.

88. The experts worked together to guide a revolution based on "responsibility to be the stewards of creativity in our communities." Creative 100, "The Memphis Manifesto: Building a Community of Ideas," 2013, http://creativeclass.com/rfcgdb/articles /manifesto.pdf.

CHAPTER 7

1. Rosabeth Moss Kanter, *Men and Women of the Corporation* (New York: Basic Books, 1977).

2. Rosabeth Moss Kanter, *The Change Masters: Innovation for Productivity in the American Corporation* (New York: Simon & Schuster, 1983), 182–205.

3. Florida, *Rise of the Creative Class*, 80.

4. Benoît Godin, "Technological Innovation: On the Origins and Development of an Inclusive Concept," *Technology and Culture* 57, no. 3 (2016): 527–556.

5. Donald Pelz and Frank Andrews, "Diversity in Research," *International Science and Technology* (July 1964): 28–36.

6. Pamela Walker Laird, *Pull: Networking and Success since Benjamin Franklin* (Cambridge, MA: Harvard University Press, 2006).

7. National Science Foundation, *Women and Minorities in Science and Engineering* (Washington, DC: NSF, 1982), 1–12.

8. Sara Jane Neustadtl, *Women Engineer* (New York: Engineers' Council for Professional Development, 1974).

9. Alice Rossi, "Women in Science: Why so Few?" *Science*, 148 (1965): 1196–1201; Edith Ruina, *Women in Science and Technology: A Report on the Workshop on Women in Science and Technology, held May 21, 22, and 23, 1973, at the Massachusetts Institute of Technology, Cambridge, Massachusetts* (Cambridge, MA: MIT Press, 1973).

10. Ruth Schwartz Cowan, "Virginia Dare to Virginia Slims: Women and Technology in American Life," Technology and Culture 20, no. 1 (1979): 51–63. Martha Moore Trescott, *Dynamos and Virgins Revisited: Women and Technological Change in History* (Lanham, MD: Scarecrow Press, 1979).

11. Sally Hacker, "Ford Fellowship Application, 1973," Sally Hacker Papers, Schlesinger Library, Radcliffe Institute, Harvard University. See also Lois Kathyrn Herr, *Women, Power, and AT&T: Winning Rights in the Workplace* (Boston, MA: Northeastern University Press, 2003).

12. Amy E. Slaton, *Race, Rigor, and Selectivity in U.S. Engineering: The History of an Occupational Color Line* (Cambridge, MA: Harvard University Press, 2010); Brian C. Odom and Stephen P. Waring, *NASA and the Long Civil Rights Movement* (Gainesville, FL: University Press of Florida, 2019); Jerry Good, "History of the National Society of Black Engineers, in *A Hammer in Their Hands: A Documentary History of Technology and the African-American Experience*, ed. Carroll W. Pursell (Cambridge, MA: MIT Press, 2005), 309–312.

13. National Science Foundation, *Women and Minorities in Science and Engineering* (Washington, DC: National Science Foundation, 1982), 1–12.

14. These programs highlighted historical successes and obstacles of Black business, including discrimination and violence, before and after emancipation. William L. Cash, Jr. and Lucy R. Oliver, eds., *Black Economic Development: Analysis and Implications, an Anthology* (Ann Arbor, MI: University of Michigan, 1975).

15. Flournoy A. Coles, Jr., "Financial Institutions and Black Entrepreneurship," *Journal of Black Studies* 3, no. 3 (1973): 329–349; Michael Ezra, ed., *The Economic Civil Rights Movement: African Americans and the Struggle for Economic Power* (New York: Routledge, 2013).

16. Leon H. Sullivan, *Build Brother Build* (Philadelphia, PA: Macrae Smith Company, 1969), 171–175; "The inequality loop: how we can help break it," *Measure*

(July 1968): 1–7; United States Community Relations Service, New Minority Enterprises: Community Development, Economic Development, Housing & Planning, Education, Legal Assistance (Washington, DC: US Department of Justice Community Relations Service, 1969; Sam Schirvar, "Manufacturing Self-Determination: Native American Industrial Development in the Postwar United States," PhD diss, (University of Pennsylvania, forthcoming 2024).

17. Rayvon Fouché, *Black Inventors in the Age of Segregation: Granville T. Woods, Lewis H. Latimer, and Shelby J. Davidson* (Baltimore, MD: Johns Hopkins University Press, 2003), 9–25.

18. Aaron E. Klein, *The Hidden Contributors: Black Scientists and Inventors in America* (New York: Doubleday, 1971); Irene Diggs, *Black Innovators* (Chicago, IL: Institute of Positive Education, 1975); The Lemelson Center, *Black Inventors and Innovators: New Perspectives* (Washington, DC: Smithsonian, 2021), 12–15. Since the 1970s, "hiddenness" has persisted as the dominant framework for Black achievement in science, technology, and innovation in dozens of books, articles, documentaries, museum exhibits, and popular films, including most recently, the book and Hollywood film *Hidden Figures*. Margot Lee Shetterly, *Hidden Figures: The American Dream and the Untold Story of the Black Women Mathematicians Who Helped Win the Space Race* (New York: William Morrow, 2016).

19. Robert C. Johnson, "Science, Technology and Black Community Development," *The Black Scholar* 15, no. 2 (1984): 32–44. See also S. E. Anderson, "Science, Technology, and Black Liberation," *The Black Scholar* 5, no. 6 (March 1974): 2–8; Sandra G. Harding, ed., *The "Racial" Economy of Science: Toward a Democratic Future* (Bloomington, IN: Indiana University Press, 1993).

20. Kanter recalled printing her own "child psychologist" business cards at age 8. Claudia Deutsch, "If at First You Don't Succeed, Believe Harder," *New York Times*, September 19, 2004, BU7.

21. Peter Cohen, *The Gospel According to the Harvard Business School* (Garden City, NY: Doubleday, 1973).

22. Rosabeth Moss Kanter, *Commitment and Community: Communes and Utopias in Sociological Perspective* (Cambridge, MA: Harvard University Press, 1972), viii, 64, back cover.

23. Kanter, *Commitment and Community*, 23, 78–80.

24. Rosabeth Moss Kanter, "Women and the Structure of Organizations: Explorations in Theory and Behavior," *Sociological Inquiry* 45, no. 2–3 (1975): 34–74. Kanter's integrative approach put her on the forefront of feminist sociology. In her coedited volume *Another Voice: Feminist Perspectives on Social Life and Social Science*, she aimed to "reassess" the social sciences in their entirety through a feminist lens. Marcia Millman and Rosabeth Moss Kanter, eds., *Another Voice: Feminist Perspectives on Social Life and Social Science* (Garden City, NY: Anchor Books, 1975), viii.

25. Kanter, *Men and Women of the Corporation*, 207.

26. Kanter, *Men and Women of the Corporation*, 166.

27. Kanter, *Men and Women of the Corporation*, xi, 103, 263.

28. Claudia M. Christie, "I'm Starting to Feel Like a Success. Why Not? I'm Just as Smart as They Are," *New England Business Profile*, March 5, 1984, 112.

29. Barry Alan Stein, "The Potential for Decentralized Community Industries," Department of Urban Studies and Planning (PhD Diss. Massachusetts Institute of Technology, 1974). Stein argued that large corporations were often less economically efficient than smaller enterprises and disastrous at matching their outcomes to community needs.

30. Rosabeth Moss Kanter and Barry Alan Stein, *A Tale of "O": On Being Different in an Organization* (New York: Harper & Row, 1980).

31. "A Tale of 'O' is no 'Tic Tac Toe," *American Rehabilitation* 7, no. 4 (1982): 1982: 26.

32. Goodmeasure, *A Tale of "O": On Being Different* (User's Manual), 1993.

33. The term came from a human relations experiment inside GM's Lordstown plant, but it aligned closely with Stein's efforts to decentralize industry. E. C. Miller, "The parallel organization structure at General Motors: An interview with Howard C. Carlson," *Personnel* 55, no. 4 (1978): 64–69.

34. Barry Alan Stein and Rosabeth Moss Kanter, "Building the Parallel Organization: Creating Mechanisms for Permanent Quality of Work life," *The Journal of Applied Behavioral Science* 16, no. 3 (1980): 371–388.

35. Kanter, *The Change Masters*, 132.

36. Grigsby herself ascended DEC's corporate ladder and eventually started her own consulting company, Next Level International. http://www.nextlevelinternational.com/management_team.html

37. Goodmeasure, Inc. "Ninety-Nine Propositions about Innovation from the Research Literature: Organizational and Individual Characteristics Affecting Development and Acceptance of Important New Practices," brochure, 1982, box 2, folder 9–23, Rosabeth Moss Kanter Papers, Baker Library, Harvard Business School. The first thirty-seven propositions were a bulleted summary of Everett Rogers's work, with latter propositions reiterating business school and techno-economics experts.

38. Kanter examined ten companies in depth. The traditional corporations included GM as well as pseudonymous firms in telecommunications, financial services, and the oil industry. Except for DEC, the "innovative" companies were happy for the publicity; they included Hewlett-Packard, Polaroid, Wang Laboratories, GE', and Honeywell.

39. Kanter, *The Change Masters*, 28, 67, 172, 281, 368.

40. In his 1981 Pulitzer Prize winning *Soul of a New Machine*, journalist Tracy Kidder lionized the invisible work of technology managers via the case of Data General manager Tom West as he motivated a team of engineers to bring a new computer to market. Tracy Kidder, *The Soul of a New Machine* (Boston, MA: Little Brown and Company, 1981).

41. Kanter also extended the identity beyond engineers and technology managers in corporate research laboratories. In the wake of *The Change Masters*, she documented over 150 subsequent studies of corporate innovators. See, for example, Gifford Pinchot III, "Introducing the 'Intrapreneur,'" *IEEE Spectrum* 22, no. 4 (1985): 74–79; Art Fry, "The Post-It Note; an Intrapreneurial Success," *Advanced Management Journal* 52 no. 3 (1987): 4–9; "3M: Organized to Innovate," *Management Review* 75 no.7 (1986): 38–39; Harold Green, "Why Intrapreneurship Doesn't Work," *Venture* 7 no. 1 (1985): 46–52.

42. Kanter, *The Change Masters*, 212, 236, 239, 306. Change masters came in many types. Some were *system builders* while others were *loss cutters* and *socially conscious pioneers*. Their character varied based on work circumstances. There were the *superstars* who succeeded regardless of their assignment, the *rescue artists* who had a reputation for saving failing initiatives, the *honeymooners* brought into an organization from outside, the *creative opportunists* who exploited oversights in stagnant companies to make change happen, and the *lone rangers* who worked against the grain of their organization. Rosabeth Moss Kanter, "Superstars and Lone Rangers Rescue Dull Enterprises," *Wall Street Journal*, January 23, 1984, 22.

43. Kanter, *The Change Masters*, 189–218.

44. Kanter, *The Change Masters*, 211, 219, 240, 355–356, 384.

45. Kanter, *The Change Masters*,159–357.

46. J. Paul Mark, *The Empire Builders: Inside the Harvard Business School* (New York: William Morrow, 1987) 87–93.

47. Rosabeth Moss Kanter, *When Giants Learn to Dance* (New York: Simon & Schuster, 1989), 45–51.

48. Charlotte Curtis, "Corporate Populist," *New York Times*, February 28, 1984, c11.

49. Cindy Skrzycki, "Harvard Business Review Names First Female Editor: Kanter Expected to Add Entrepreneurial Spirit," *Washington Post*, December 1, 1989, F1; Susan McHenry, "Rosabeth Moss Kanter," *Ms. Magazine* profile 13, no. 7 (1985): 62–63, 107.

50. Lawrence C. Soley, *Leasing the Ivory Tower: The Corporate Takeover of Academia* (Boston, MA: South End Press, 1995), 79–80; Maureen McFadden, "The Guru of the Corporation," *Working Woman* 11 (1986): 60–66.

51. Rosabeth Moss Kanter, John Kao, Fred Wiersema, *Innovation: Breakthrough Ideas at 3M, Dupont, GE, Pfizer, and Rubbermaid* (New York: HarperCollins, 1997).

52. Rosabeth Moss Kanter, *Evolve! Succeeding in the Digital Culture of Tomorrow* (Boston, MA: Harvard Business School Press, 2001), 3.

53. In a 1993 edition of her book, Kanter documented the changing demographics as well as qualitative shifts inside Indsco. Rosabeth Moss Kanter, *Men and Women of the Corporation* (New York: Basic Books, 1993), 289–330.

54. The National Foundation for Women Business Owners, *Paths to Entrepreneurship: New Directions for Women in Business* (April 1998), i; Timothy D. Boston, "Trends in Minority-Owned Businesses," in *America Becoming: Racial Trends and Their Consequences: Volume II*, ed. Neil J. Smelser, William Julius Wilson, and Faith Mitchell (Washington, DC: The National Academies Press, 2001), 190–221.

55. The region's foreign-born population doubled between 1975 and 1990 and expanded rapidly after the 1990 Immigration and Nationality Act, which nearly tripled the number of H-1B visa holders. High-tech migrants established dozens of associations that served as mutual aid networks such as the Asia America Multitechnology Association and the Silicon Valley Indian Professionals Association. AnnaLee Saxenian, *The New Argonauts: Regional Advantage in a Global Economy* (Cambridge, MA: Harvard University Press, 2006), 49–51, 341–346; Shalini Shankar, *Desi Land: Teen Culture, Class, and Success in Silicon Valley* (Durham, NC: Duke University Press, 2008).

56. Joan Acker, "Theorizing Gender, Race, and Class in Organizations," in *Handbook of Gender, Work, and Organization*, ed. Emma L Jeans, David Knights, and Patricia Yancey Martin (West Sussex, UK: John Wiley & Sons, 2011), 65–80.

57. Judy Wajcman, *Managing Like a Man: Women and Men in Corporate Management* (Cambridge, UK: Polity, 1998), 46, 161; Janice D. Yoder, "Rethinking Tokenism: Looking Beyond Numbers," *Gender & Society* 5, no. 2 (1991): 178–192.

58. Joan Rothchild, ed., *Women, Technology, and Innovation* (Oxford, UK: Pergamon Press, 1982).

59. National Research Council, *Women Scientists and Engineers Employed in Industry: Why So Few?* (Washington, DC: National Academies Press, 1994), 69.

60. See, for example, Sondra Thiederman, *Bridging Cultural Barriers for Corporate Success: How to Manage the Multicultural Work Force* (Lexington, MA: Lexington Books, 1991); Odette Pollar and Rafael Gonzalez, Jr. *Dynamics of Diversity: Strategic Programs for Your Organization* (Menlo Park, CA: Crisp Publications, Inc. 1994); Elsie Y. Cross, Judith H. Katz, Frederick A. Miller, and Edith W. Seashore, *The Promise of Diversity: Over 40 Voices Discuss Strategies for Eliminating Discrimination in Organizations* (Burr Ridge, IL: Irwin, 1994).

61. R. Roosevelt Thomas, Jr. "From Affirmative Action to Affirming Diversity," *Harvard Business Review* 68, no.2 (1990): 107–117; Mary Gentile, ed. *Differences that Work: Organizational Excellence Through Diversity* (Boston, MA: Harvard Business Review, 1994); Roosevelt R. Thomas, Jr. *Beyond Race and Gender: Unleashing the Power of Your Total Workforce by Managing Diversity* (New York: AMACOM, 1991).

62. The former Stanford affirmative action officer was interviewed in nearly every article about high-tech diversity in the 1980s. Carmela C. Mellado, Ray Mellado, and Frank Armendariz, "Apple's New Approaches to Affirmative Action," *Hispanic Engineer* 5, no. 1 (1989): 32–34, 36; Tim Larimer, "It's still Anglo at the Top," *Mercury News*, October 1, 1989, 1A; Joseph Rodriguez, "It's No More Mr. White Guy for Silicon Valley, The Multiculture Future of America Is Already Here" Top," *Mercury News*, September 29, 1991, 7C.

63. Lorretta M. Green, Sharon Skeeter, and Constance Mitchel, "What's Cooking in America's High-Tech Hot Spots," *Black Enterprise* 15, no. 11 (1985): 257–264; Grady Wells, "Managing 3M's Diversity," *US Black Engineer* 12, no. 5 (1982), 26–32.

64. Ken Smikle, "Engineering a Magazine Career," *Black Enterprise* 17, no. 6 (January 1987): 23; Tyrone Taborn, *The HistoryMakers*, interviewed by Cheryl Butler, July 28, 2007.

65. Florida, *Rise of the Creative Class*, 79.

66. David A. Thomas and Suzy Wetlaufer, "A Question of Color: A Debate on Race in the U.S. Workplace," *Harvard Business Review* 75, no. 5 (1997): 118–132.

67. Michael S. Dukakis and Rosabeth Moss Kanter, *Creating the Future: The Massachusetts Comeback and Its Promise for America* (New York: Summit Books, 1988), 29–39, 56.

68. A very early conceptualization appeared in Peter J. Berger and Richard John Neuhaus, *To Empower People* (Washington, DC: American Enterprise Institute, 1977).

69. NCNE's initiatives included experimental programs in which residents of public housing projects managed their own communities. Robert L. Woodson, Larry F Crowe, Scott Stearns, and HistoryMakers (Video oral history collection), dirs. 2016. *The Historymakers Video Oral History with Robert Woodson*. HistoryMakers; Robert Weissberg, *The Politics of Empowerment* (Westport, CT: Praeger, 1999); Clint Bolick, *Transformation: The Promise and Politics of Empowerment* (Oakland, CA: ICS Press, 1998).

70. Paul Pryde, "Investing in People: A New Approach to Job Creation," in *On the Road to Economic Freedom: An Agenda for Black Progress* ed. Robert L. Woodson (Washington, DC: Regnery Gateway, 1987), 27–44. Quote on 44.

71. Burt Solomon, "Power to the People?" *National Journal*, January 26, 1991, 206–209; James P. Pinkerton, *What Comes Next: The End of Big Government—And the New Paradigm Ahead* (New York: Hyperion, 1995).

72. Gerald F. Seib, "Modem Operandi: As Politics Go Digital, Info Highway Traffic Is Heavy in Right Lane," *Wall Street Journal*, March 16, 1994, A1.

73. David Osborne and Ted Gaebler, *Reinventing Government: How the Entrepreneurial Spirit is Transforming the Public Sector* (Reading, MA: Addison-Wesley, 1992), n1, 19, 269–270.

74. Benjamin F. Chavis, "Support Brother Espy," *Civil Rights Journal*, March 4, 1993, A6.

75. Bill Clinton, "State of the Union Address," January 24, 1995. https://millercenter
.org/the-presidency/presidential-speeches/january-24-1995-state-union-address; Bill
Clinton, "Address on Affirmative Action," July 19, 1995, https://millercenter.org/the
-presidency/presidential-speeches/july-19-1995-address-affirmative-action. On welfare reform, Clinton argued for a shift from "entitlement programs to empowerment
programs," Bill Clinton, "Address Before a Joint Session of Congress," February 17,
1993, https://millercenter.org/the-presidency/presidential-speeches/february-17
-1993-address-joint-session-congress.

76. The idea of enterprise zones stemmed from British urban planning in the late
1970s to encourage industrial innovation by selectively eliminating regulations and
taxes in impoverished urban communities with the hope of fostering both low skill
labor and small firm technology entrepreneurship. Peter Hall, "Enterprise zones:
a justification," *International Journal of Urban and Regional Research* 6, no. 3 (1982):
416–421; Marc Bendick and David Rasmussen, "Enterprise Zones and Inner City Economic Revitalization," in *Reagan and the Cities*, ed. George Peterson and Carol Lewis
(Washington, DC: The Urban Institute, 1986); Wilton Hyman, "Empowerment Zones,
Enterprise Communities, Black Business, and Unemployment," *Washington University
Journal of Urban and Contemporary Law* 53 (1998): 143–170.; Timothy Mason, *Bank-
ing on Black Enterprise: The Potential of Emerging Firms for Revitalizing Urban Economies*
(Washington, DC: Joint Center for Political and Economic Studies, 1993).

77. Ron Brown was the first Black head of the Department of Commerce, who died
tragically in 1996 when his plane crashed in Croatia. IBM was one of the first winners
for its Diversity Councils and Network Groups, which had led to gains in the percentage of minorities in women in executive positions. Deborah Parkinson, *Corporate
Achievement in Employee and Community Relations: The Ron Brown Award for Corporate
Leadership* (USA: The Conference Board, Inc. 1999), 47–49; John T. Barber, *The Black
Digital Elite: African American Leaders of the Information Revolution* (Westport, CT: Praeger, 2006), 31–38.

78. US National Telecommunications and Information Administration, *Falling
Through the Net: A Survey of "Have Nots" in Rural and Urban* America (National Telecommunications and Information Administration, 1995), https://www.ntia.gov
/page/falling-through-net-survey-have-nots-rural-and-urban-america; US National
Telecommunications and Information Administration, *Falling through the Net: Defining the Digital Divide* (Washington, DC: US Department of Commerce, 1999); Benjamin M. Compaine, *The Digital Divide Facing a Crisis or Creating a Myth?* (Cambridge,
MA: MIT Press, 2001); Logan Hill, "Beyond Access: Race, Technology, Community,"
in *Technicolor: Race, Technology, and Everyday Life*, ed. Alondra Nelson and Thuy
Linh N. Tu (New York: NYU Press, 2001): 13–33.

79. Julia Angwin and Laura Castaneda, "The Digital Divide: High Tech Boom a Bust
for Blacks, Latinos," *San Francisco Chronicle*, May 4, 1998, https://www.sfgate.com

/news/article/The-Digital-Divide-High-tech-boom-a-bust-for-3007911.php. See also Joel Dreyfuss, "Valley of Denial," *Fortune*, July 19, 1999, 60–61.

80. Jeordan Legon, "Latinos Caught in Tech Gap Revolution Fails to Even Playing Field," *Mercury News*, April 7, 1996, A1; Edward J. W. Park, "Racial Ideology and Hiring Decisions in Silicon Valley," *Qualitative Sociology* 22, no. 3 (1999): 223–233.

81. "Jesse Jackson: Racism in Cyberspace," *Sun*, March 4, 1999, 1; Roger O. Crockett, et. al. "Jessie's New Target: Silicon Valley," *Business Week*, July 12, 1999, 111–112.

82. Mark Hanauer, "A Friendly Frontier for Female Pioneers," *Fortune*, June 25, 1984, 78–85. Profiled were Sandra Kurtzig, Lorraine Mecca, Therese Myers, Shirley Eis, Stina Hans, Ann Piestrup, Ann Winblad, Margaret Hamilton, and Saydean Zeldin. Mecca's claim was wildly inaccurate. There are no good indices at the time, but a recent critique of the venture capital industry shows that as late as 2016 only 11 percent of investment partners were women. NVCA, Venture Forward, and Deloitte, *NVCA Human Capital Survey*, 3rd ed., March 2021, https://nvca.org/wp-content/uploads/2022/08/vc-human-capital-survey-3rd-edition-2021.pdf

83. Barbara McIntosh, "The Reprograming of Sandra Kurtzig," *San Jose Mercury News*, December 21, 1986. Leslie Berlin, *Troublemakers: Silicon Valley's Coming of Age* (New York: Simon & Schuster, 2017), 75–88.

84. Emily T. Smith, Alice LaPlante, Paul Angiolillo, and Catherine L. Cantrell, "The Women Who Are Scaling High Tech's Heights," *Business Week*, August 28, 1989, 86–88.

85. For one of the very few mainstream profiles in the 1980s, see Rudolph A. Pyatt Jr., "Black Entrepreneurs Find Success in High-Tech," *Washington Post*, September 15, 1986, WB1–3.

86. "30 New Leaders for the Eighties," *Black Enterprise* 11, no. 8 (1981): 33–38.

87. "Business Innovator of the Year," *Black Enterprise* 27, no. 4 (1996): 76–77; Yvonne R. Keith, "1995 Black Engineer of the Year Awards Winners," *US Black Engineer* 19, no. 1 (1995): 41–57. Men made up most innovators in these publications. Elsewhere, NASA data scientist Valerie Thomas highlighted Black women pioneers including entrepreneur Zella Jackson and astronaut Mae Jemison. Valerie L. Thomas, "Black Women Engineers and Technologists," *SAGE* 6 (1989), 24–32.

88. Britt Robson, "Intrapreneurship" Making Those Inside Moves," *Black Enterprise* 15, no. 11 (1985): 197–202. For similar narratives see: "Winning Success in Silicon Valley," *Ebony* 40, no. 1 (1984): 37–40; Tyrone Taborn and Grady Wells, "Sam T. Sims of General Motors C-P-C," *US Black Engineer* (Spring 1986): 54–55, 82–84; Carmella Mellado, "Jim E. Tarro, Plant Quality Manager for Digital Equipment Corporation in Albuquerque," *Hispanic Engineer* 4, no. 1 (1988): 22–26.

89. Tariq K. Muhammad and Cheryl Coward, "The Black Digerati," *Black Enterprise* 18, no. 8 (1998): 49–54.

90. Roger O. Crockett, "Invisible-and Loving It" *Business Week*, October 5, 1998, 124–128.

91. Tyrone D. Taborn, "High-tech's 'Invisible Man,'" *US Black Engineer & Information Technology* 25, no. 5 (2002): 14.

92. Barber, *The Black Digital Elite*; Roy L. Clay, Sr. and M. H. Jackson. *Unstoppable: The Unlikely Story of a Silicon Valley Godfather* (Oakland, CA: RLC Publishing, 2022). For recent examples, see Kathy Cotton, *A Place at the Table*, 2018, https://www .kathycottondigitalstoryteller.com; Jessica Guynn, "The Race to Save Silicon Valley's Untold Black History," *USA Today*, June 12, 2023, https://www.usatoday.com/story /money/2023/06/07/silicon-valley-tech-black-history-roy-clay/70262081007/.

93. "Booz Allen Hamilton," *US Black Engineer* 19, no. 1 (1995), 5.

94. "Johnson and Johnson," *US Black Engineer* 18, no. 1 (1994), 87. Mobil used a similar historical approach to tell the up-from-poverty life history of employee Arnold Stancell, a 1992 Black Engineer of the Year award recipient. Mobil, "A Future Worth the Dream," *Black Enterprise* 16, no. 1 (1992), 107.

95. "Advanced Micro Devices," *US Black Engineer* 18, no. 3 (1994), 31.

96. "Innovators," *Black Enterprise* 19, no. 4 (1988): 23.

97. "Windows 98," *New York Times*, May 24, 1998, 12.

98. Scott E. Page, *The Difference: How the Power of Diversity Creates Better Groups, Firms, and Societies* (Princeton, NJ: Princeton University Press, 2007); Cedric Herring, "Does Diversity Pay? Race, Gender, and the Business Case for Diversity," *American Sociological Review* 74, no. 2 (2009): 208–224; Sylvia Ann Hewlett, Melinda Marshall, and Laura Sherbin, "How Diversity Can Drive Innovation," *Harvard Business Review* 91, no. 12 (2013): 30.

99. In 2017, Facebook retired it mission for the value-neutral "bring the world closer together." Khari Johnson, "Facebook gives up on making the world more open and connected, now wants to bring the world closer together," *Venture Beat*, June 22, 2017, https://venturebeat.com/social/facebook-gives-up-on-making-the-world-more -open-and-connected-now-wants-to-bring-the-world-closer-together/

100. Adams Nager, David Hart, Stephen Ezell, and Robert D. Atkinson, *The Demographics of Innovation in the United States*, Information Technology and Innovation Foundation, February 24, 2016, 8, 20.

101. Buck Gee, Denise Peck, and Janet Wong, "Hidden in Plain Sight: Asian American Leaders in Silicon Valley," 2015, https://asiasociety.org/sites/default/files/inline -files/HiddenInPlainSight_Paper_042.pdf.

102. See chapter 9 for an expanded discussion of inclusive innovation programs.

103. Lucinda M. Sanders and Catherine Ashcraft, "Confronting the Absence of Women in Technology Innovation," in Wisnioski, Hintz, and Kleine, *Does America*

Need More Innovators?, 323–343. See also, Gry Agnete Alsos, Elisabet Ljunggren, and Ulla Hytti, "Gender and Innovation: State of the Art and a Research Agenda," *International Journal of Gender and Entrepreneurship* 5 (2013): 236–256.

104. Sheryl Sandberg, *Lean In: Women, Work, and the Will to Lead* (New York: Alfred A. Knopf, 2013).

105. bell hooks, "Dig Deep: Beyond Lean In," *The Feminist Wire*, October 28, 2013, http://www.thefeministwire.com/2013/10/17973/.

106. FemTechNet, *manifesto*, https://www.femtechnet.org/publications/manifesto/.

CHAPTER 8

1. "National Museum of American History, Innovative Lives: Classroom Enrichment, Grades 5–9," 1998, box 1, Advisory Board, Accession 3–117, Lemelson Center, Administrative Records, Smithsonian Institution Archives (hereafter cited as LemCen).

2. "Excerpts from Interviews with Inventors for *Invention at Play*," box 10, IAP Inventor Interviews, Accession 12–371, LemCen 12–371.

3. "Playful Inventing Activities Cart," September 2003, box 10, PIE, LemCen 12–371.

4. Norman Brosterman, *Inventing Kindergarten* (New York: Harry N. Abrams, 1997).

5. Patricia Albjerg Graham, *Schooling America: How the Public Schools Meet the Nation's Changing Needs* (New York: Oxford, 2005); Johann N. Neem, *Democracy's Schools: The Rise of Public Education in America* (Baltimore, MD: Johns Hopkins University Press, 2017).

6. Jamie Cohen-Cole, *The Open Mind: Cold War Politics and the Science of Human Nature* (Chicago, IL: University of Chicago Press, 2013); Amy F. Ogata, *Designing the Creative Child: Playthings and Places in Midcentury America* (Minneapolis, MN: University of Minnesota, 2013); John L. Rudolph, *How We Teach Science: What's Changed, and Why It Matters* (Cambridge, MA: Harvard University Press, 2019); Samuel Weil Franklin, *The Cult of Creativity: A Surprisingly Recent History* (Chicago, IL: University of Chicago Press, 2023).

7. Commerce Technical Advisory Board, *Recommendations on Learning Environments for Innovation* (Washington, DC: US Department of Commerce, 1980), 13–14, 43.

8. National Commission on Excellence in Education. *A Nation at Risk: The Imperative for Educational Reform* (Washington, DC: United States Government, 1983).

9. US Patent and Trademark Office, *The Inventive Thinking Curriculum Project* (Washington, DC: Department of Commerce, 1991).

10. US Patent and Trademark Office, *The Inventive Thinking Curriculum Project*; Donald J. Quigg, "Technology on the Move: The Role of Patents," in *Inventive Minds:*

Creativity in Technology, ed. Robert R. Weber and David N. Perkins (New York: Oxford University Press, 1992), 311–316.

11. Bettijane Levine, "Inventing Is Kid's Stuff, Too," *Los Angeles Times*, November 1, 1995, E1; Susan Trausch, "In DC, Little Inventors, Big Ideas," *Boston Globe*, January 10, 1987, 1, 9; Susan Gervasi, "The Children of Invention," *Washington Post*, February 12, 1991, D5. Cañedo would go on to become superintendent of Buffalo Public Schools, overseeing a tumultuous transformation to charter schools amid severe budget crises in the struggling industrial city.

12. US Patent and Trademark Office, *The Inventive Thinking Curriculum Project*.

13. Project XL, *Black Innovators in Technology: Inspiring a New Generation* (Washington, DC: US Patent and Trademark Office, 1990). Lisbeth Gant Stevenson, *African-American History: Heroes in Hardship* (Cambridge, MA: Cambridgeport Press, 1992), 186–201; James Michael Brodie, *Created Equal: The Lives and Ideas of Black American Innovators* (New York: William Morrow, 1993); and C. R. Gibbs. *Black Inventors: From Africa to America, Two Million Years of Invention and Innovation* (Silver Spring, MD: Three Dimensional Publishing, 1995.

14. N. E. Collins, *Catalog of Programs, Activities, and Assistance Materials for Young Inventors* (Lemont, IL: Argonne National Laboratories, 1990); Ellen Harvey Showell and Fred M. B. Amram, *From Indian Corn to Outer Space: Women Invent in America.* Peterborough, NH: Cobblestone Publishing, 1995), 146–152.

15. Academy of Applied Sciences, "I Was Meant to Invent: Student Handbook," 1998, box 2, New Hampshire Young Inventors, LemCen 08–038.

16. First Robotics, "Havoc on the Hill: 1998 First Capitol Hill Invitational," June 18, 1998, box 1, Dean Kamen, LemCen 12–582.

17. Annette Bernhard, "Getting Big Ideas," *The Dallas Morning News*, January 8, 1987, 25A; Associated Press, "'Tall Hanger' Inspiration for Inventors' Campaign," *Arlington Heights Daily Herald*, January 10, 1987, 83.

18. Seymour Papert, *Mindstorms: Children, Computers, and Powerful Ideas* (New York: Basic Books, 1980), 177–178, 186, 209.

19. Idit Harel and Seymour Papert, "Software Design as a Learning Environment," *Interactive Learning Environments* 1 (1990): 1–32.

20. Jeanne Bamberger, "Action Knowledge and Symbolic Knowledge: The Computer as Mediator," in *High Technology and Low-Income Communities: Prospects for the Positive Use of Advanced Information Technology*, ed. Donald Schön, Bish Sandal, and William J. Mitchell (Cambridge, MA: MIT Press, 1999): 237–261

21. Mitch Resnick, Stephen Ocko, and Seymour Papert, "LEGO, Logo, and Design," *Children's Environments Quarterly* 5:4 (1988): 4–18; Mark Pesce, *The Playful World: How Technology is Transforming Our Imagination* (New York: Ballentine Books, 2000), 82–97; 123–125.

22. Natalie Rusk, "Designing Learning Environments that Engage Young People as Creators," in Wisnioski, Hintz, and Kleine, *Does America Need More Innovators?* 281–298.

23. Mitch Resnick, *Lifelong Kindergarten: Cultivating Creativity through Projects, Passion, Peers, and Play* (Cambridge, MA: MIT Press, 2017), 7–10.

24. These philanthropies established patterns for grant-making and scientific research as they promoted the benefits of industrial capitalism. They identified their ideal recipient as the "exceptional man" dedicated to the creation and adoption of new ideas who would otherwise be unable to succeed without foundation support. The model proved so successful that the US government modeled NSF on private philanthropic practices. Robert Kohler, *Partners in Science: Foundations and Natural Scientists, 1900–1945* (Chicago, IL: University of Chicago Press, 1991); David E. Weischadle, "The Carnegie Corporation and the Shaping of American Educational Policy," in *Philanthropy and Cultural Imperialism: The Foundations at Home and Abroad*, ed. Robert F. Arnove (Boston, MA: G.K. Hall & Co. 1990), 363–384. See also the discussion of Gardner in chapter 2.

25. Rakesh Khurana, Kenneth Kimura, and Marion Fourcade, *How Foundations Think: The Ford Foundation as a Dominating Institution in the Field of American Business Schools* (Boston, MA: Harvard Business School, 2011).

26. John Gardner took on the role of statesman and defender of robust federal investment and engagement in social programs against Reagan administration efforts to use philanthropy and volunteerism as a reason to reduce federal programs. "President's Task Force on Private Sector Initiatives," Gardner, John 92–054, box 2, Reagan Task Force on Private Sector Initiatives, John W. Gardner Papers, M0659, Stanford University Special Collections and University Archives.

27. Fran Smith, "Community Relations 101: Dot-com Do-Gooders Try to Win Over Their Angry Neighbors," *Red Herring*, December 4, 2000, 256–267; Douglas Foster, "Paying the Price," *GROK* (October 2000): 66–73; John Pepin, "Venture Capitalists and Entrepreneurs Become Venture Philanthropists," *Journal of Philanthropy and Marketing* 10, no. 3 (2005): 165–173.

28. Anne Morgan, *Prescription for Success: The Life and Values of Ewing Marion Kauffman* (Kansas City, MO: Andrews and McMeel, 1995), 319–360.

29. Lawrence D. Maloney, "Lone Wolf of the Sierras," *Design News*, March 6, 1995, 70–84.

30. Hintz, *American Independent Inventors*, 233–235; Nicholas Varchaver, "The Patent King," *Fortune*, May 14, 2002, 202–216; Henry Petroski, "An Independent Inventor," *American Scientist* 86 (May–June) 1998: 222–225; George Goldberg and Rick Morgan, "Lemelson Receives Go-Ahead for Patent Infringement Suit," *SCAN: The Data Capture Report*, May 30, 1997, 1–3.

31. Thurow was the author of MIT's *Made in America* study described in chapter 6.

32. "What does the foundation do?" box 1, Indonesia Presentations, LemCen 08–038.

33. Petroski, "An Independent Inventor," 222–225.

34. Phil Weilerstein "Dear Colleague," August 15, 1999, box 3, NCIIA 1999b LemCen 03–117.

35. The University Innovation Fellows later became a subsidiary program of Stanford's d.school.

36. Initiated in 1995 at the start of the public internet, NCIIA's undergraduate inventors also launched multimedia ventures that combined digital technology with arts, humanities, and business such as a pre-Google "internet secretary" that applied AI to web browsing and a digital humanities project on Black writing. "Inventions Past, Present, and Future," March 11, 1999, box 3, NCIIA Events; "Lemelson National Program at Hampshire College: Invention, Innovation and Creativity, Report of Activities 1993–1994, Hampshire College," box 2, Hampshire 1994, LemCen 03–117.

37. By the early 2000s, Lemelson support amounted to $40 million, permanently endowing the Center.

38. Henry Ford Museum & Greenfield Village, *1991 Annual Report*, box 2, Henry Ford, LemCen 08–038.

39. Portia P. James, *The Real McCoy: African-American Invention and Innovation, 1619–1930* (Washington, DC: Smithsonian Institution Press, 1990).

40. "S.J. Tech Museum may downsize," *San Jose Mercury News*, May 13, 1993, A1–16A.

41. "Creating Tomorrow's Innovators: Challenge Activities Building on Science Education-Industrial Partnerships," box 228, Tech Museum of Innovation General 1992–1993, William Hewlett Papers, M1995, Stanford University Special Collections and University Archives.

42. "Meeting with Judith Fritz, Vice President of Programs at the Tech Museum of Innovation," February 18, 1999, box 1, Traveling Exhibition, LemCen 12–371. On the Tech's Black innovator programming see Ben Stocking, "Tech Museum Focuses on Black Innovators, Minority Kids Find Role Models," *Mercury News*, March 21, 1988, 1B.

43. Arthur P. Molella, "Budget Cuts from a Cultural Perspective," July 7, 1999, box 4, Culture of Budget Cuts, LemCen 12–582. Robert C. Post and Arthur P. Molella, "The Call for Stories at the Smithsonian Institution: History of Technology and Science in Crisis," *ICON* 3 (1997): 44–82.

44. The Hands On Science Center was part of the Science in American Life initiative, a major exhibit to merge history and engagement with contemporary science. It was one of the first Smithsonian corporate funded events, supported by Merck, Monsanto, and the American Chemical Society. Gordon L. Nelson to Arthur

Molella, March 1, 1998, box 1, American Chemical Society, 1988; Spencer Crew to John R. Taylor, July 7, 1998, box 3, Merck funding records, 1998–2006, LemCen 12–582.

45. Joanna Hanes-Lahr to Arthur Molella, December 16, 1993, box 2, Lemelson Foundation Funding; Donna Ari to Art Molella, July 15, 1994, LemCen 12–582.

46. The Smithsonian's pitch emphasized "innovation in science and technology," which it later crossed out for just "innovation," and then jostled between "innovation" or "invention" as the lead. "Draft: Program Goals," 1994, Proposal Development 1993–1994, box 1, LemCen 03–115.

47. National Museum of American History, *Innovative Lives: Classroom Enrichment, Grades 5–9*, 1998, box 1, Advisory Board LemCen 03–117.

48. A-MAN was founded by Black aerospace engineer Hal Walker, Jr., the inventor of the laser that NASA placed on the moon during the Apollo 11 mission to take accurate geospatial measurements.

49. Andrew J. Pekarik and Abigail J. Dreibelbis, "Investigating Invention: Results from the 1999 Invention Interview Study," January 2000, box 1, Advisory Meeting Mailing, LemCen 3–117.

50. Gretchen Jennings, "Time to Listen," *Curator* 46/4 (October 2003): 371–384.

51. This movement built on earlier successes of San Francisco's Exploratorium and Brooklyn's Children's Museum, which fused science, arts, and play in pursuit of creativity established models for exhibition design in which children reimagined the future. Ogata, *Designing the Creative Child*, 175–186; Rebecca Onion, *Innocent Experiments: Childhood and the Culture of Science in the United States* (Chapel Hill, NC: University of North Carolina Press, 2016).

52. Mihaly Csikszentmihalyi, for example, identified innovators as those in a society who "do not play by the usual rules." Mihaly Csikszentmihalyi, *Creativity: Flow and the Psychology of Discovery and Invention* (New York: Harper Perennial, 1996), 72, 106, 163, 197–198, 208, 253.

53. Robert Scott Root-Bernstein, Maurine Bernstein, and Helen Garnier, "Correlations Between Avocations, Scientific Style, Work Habits, and Professional Impact of Scientists," *Creativity Research Journal* 8 (1995).

54. Alison Gopnik, Andrew N. Meltzoff, and Patricia K. Kohl, *The Scientist in the Crib: Minds, Brains, and How Children Learn* (New York: William Morrow, 1999).

55. Lemelson Center, *Invention at Play*.

56. NSF Proposal Number 0125417 "Invention at Play," box 7, NSF, LemCen 120371.

57. The PIE Network collaborated with hands-on museums across the country, including the Science Museum of Minnesota, Exploratorium, Lemelson Center,

American Visionary Art Museum, Fort Worth Museum of Science and History, and the MIT Museum.

58. Lemelson Center for the Study of Invention and Innovation, *Invention at Play: Educator's Manual*, https://invention.si.edu/sites/default/files/Invention-at-Play-Educators-Manual.pdf; "Video: Play to the Future," *Invention at Play*, https://invention.si.edu/tags/invention-play.

59. Funded by Draper Laboratories and the Ford Motor Company, Spark!Lab remains one of the NMAH's most frequented attractions at the Smithsonian and nationwide via satellite centers from Holland, Michigan, to Anchorage, Alaska.

60. Monica M. Smith, "Invention at Play: An Award-winning Traveling Exhibition," in *Museums at Play: Games, Interaction and Learning*, ed. Katy Beale (Edinburgh, UK: Museums Etc., 2011), 440–58; Monica M. Smith, "Playful Invention, Inventive Play," *International Journal of Play* 5, no. 3 (2016): 244–261.

61. "Play" appears nearly 150 times in the book. Robert Scott and Michelle Root-Bernstein, *Sparks of Genius: The 13 Thinking Tools of the World's Most Creative People* (Boston, MA: Houghton Mifflin, 1999).

62. Daniel H. Pink, *A Whole New Mind: Why Right-brainers Will Rule the Future* (New York: Riverhead Books, 2005), 2.

63. "Imagination's Playground," *Wall Street Journal*, October 18, 1996, B14. The television program *Nightline* highlighted IDEO's fun-loving approach as it went about reimagining the grocery shopping experience, the results of which were on display. "The Deep Dive," *Nightline* 02-09-99, ABC News Home Video N990209–01.

64. Tom Kelley, with Jonathan Littman, *The Art of Innovation: Lessons in Creativity from IDEO, America's Leading Design Firm* (New York: Currency, 2001), 20, 73, 94, 129, 279.

65. Michael Schrage, *Serious Play: How the World's Best Companies Simulate to Innovate* (Boston, MA: Harvard Business School Press, 2000), 2.

66. Dale Russakoff, "Mind Games for Tech Success: You've got to Play to Win." *Washington Post*, May 8, 2000, A1, A4.

67. Bruce Sterling, "Greetings from Burning Man!" *Wired*, November 1, 1996, https://www.wired.com/1996/11/burningman-2/.

68. Pink, *A Whole New Mind*, 61.

69. Sir. Ken Robinson, *Do Schools Kill Creativity?* [Video] February 2006. https://www.ted.com/talks/sir_ken_robinson_do_schools_kill_creativity.

70. Dale Dougherty, "The Making of Make," *Make* 1 (2005): 7.

71. Saul Griffith, "The Playful Scientist," *Make* 4 (2005): 46–47.

72. "Take the 'Mayor Maker Challenge," May 12, 2014, https://obamawhitehouse.archives.gov/blog/2014/05/15/challenging-mayors-help-make-difference.

73. Christina Vercelletto, "Raise the Next Steve Jobs," *Parenting* 260 (February 2012): 68–73.

74. Jan Burns, *Shigeru Miyamoto: Nintendo Game Designer* (Detroit, MI: Kidhaven Press, 2006).

75. Jake Cook, "IDEO: Big Innovation Lives Right on the Edge of Ridiculous Ideas," *The Behance Blog*, September 14, 2011, https://www.behance.net/blog/ideo-big -innovation-lives-right-on-the-edge-of-ridiculous-ideas.

CHAPTER 9

1. National Science Foundation, *Empowering the Nation through Discovery and Innovation: NSF Strategic Plan for Fiscal Years 2011–2016* (Arlington, VA: National Science Foundation, 2011).

2. Marc Philippe Eskenazi, "3 Under 3," *The New Yorker*, 2014, http://www .newyorker.com/humor/daily-shouts/3-under-3.

3. "Internet/Broadband Fact Sheet," Pew Research Center, November 16, 2022, https://www.pewresearch.org/internet/fact-sheet/internet-broadband/#smartphone -dependency-over-time; Felix Richter, "How Facebook Grew from 0 to 2.3 Billion Users in 15 Years," World Economic Forum, February 5, 2019, https://www.weforum .org/agenda/2019/02/how-facebook-grew-from-0-to-2-3-billion-users-in-15-years/.

4. "The Innovators Issue," *The New Yorker*, May 14, 2007.

5. Malcolm Gladwell, *The Tipping Point: How Little Things Can Make a Big Difference* (New York: Little, Brown and Company, 2000); Malcolm Gladwell, *Outliers: The Story of Success* (New York: Little, Brown and Company, 2008).

6. Ben Mezrich, *The Accidental Billionaires: The Founding of Facebook: A Tale of Sex, Money, Genius and Betrayal* (New York: Doubleday, 2009); Brad Stone, *The Everything Store Jeff Bezos and the Age of Amazon* (New York: Little, Brown and Company, 2013); Phil Knight, *Shoe Dog* (New York: Scribner, 2016); Ashlee Vance, *Elon Musk: Tesla, SpaceX, and the Quest for a Fantastic Future* (New York: HarperCollins, 2015).

7. Ed Catmull, and Amy Wallace, *Creativity, Inc* (New York: Random House, 2014); Josh Eells, "Dr. Dre and Jimmy Iovine's School for Innovation," *The Wall Street Journal*, November 5, 2014, https://www.wsj.com/articles/dr-dre-and-jimmy-iovines -school-for-innovation-1415238722.

8. Barry Blitt, "Book of Life," *The New Yorker*, October 17, 2011, cover.

9. Isaacson, *Steve Jobs*, 566–567. Isaacson followed up with *The Innovators*, a history of computing that cast "inventors, hackers, geniuses, and geeks" as leading agents of American progress. Walter Isaacson, *The Innovators: How a Group of Inventors, Hackers, Geniuses and Geeks Created the Digital Revolution* (New York: Simon & Schuster, 2015).

10. Macon Phillips, "President Barack Obama's Inaugural Address," National Archives and Records Administration, January 21, 2009, https://obamawhitehouse.archives.gov/blog/2009/01/21/president-Barack-obamas-inaugural-address.

11. Maggie Jo Buchanan et al., "President Barack Obama: The Entrepreneur-in-Chief?" Center for American Progress, November 30, 2012, https://www.americanprogress.org/article/president-barack-obama-the-entrepreneur-in-chief/; Jon Gertner, "Inside Obama's Stealth Startup," *Fast Company*, June 15, 2015, https://www.fastcompany.com/3046756/obama-and-his-geeks; Pavithra Mohan, "7 Tech CEOS and Execs on Why Obama Was Innovator in Chief—Fast Company," *Fast Company*, January 20, 2017, https://www.fastcompany.com/3067375/7-tech-ceos-and-execs-on-why-obama-was-innovator-in-chief; Robert Safian, "President Obama: The Fast Company Interview," *Fast Company*, June 15, 2015, https://www.fastcompany.com/3046757/president-barack-obama-on-what-we-the-people-means-in-the-21st-century; Evan Ratcliff, "The Wired Presidency: Can Obama Really Reboot the White House?," *Wired*, January 19, 2009, https://www.wired.com/2009/01/ff-obama/.

12. White House, "President Obama Lays Out Strategy for American Innovation," https://obamawhitehouse.archives.gov/the-press-office/president-obama-lays-out-strategy-american-innovation.

13. Elizabeth Segran, "Meet the Geeks: The DC Tech Corps's Leading Edge," *Fast Company*, June 15, 2015, https://www.fastcompany.com/3046985/meet-the-geeks-the-dc-tech-corps-leading-edge.

14. From 1985 to 2007, there were five women recipients of the National Medal of Technology and Innovation, and only two as individual winners, representing less than 3 percent. During the Obama administration, women accounted for eight individual winners, or 19 percent of awardees. USTPO, "National Medal of Technology Recipients," USPTO, https://www.uspto.gov/learning-and-resources/ip-programs-and-awards/national-medal-technology-and-innovation/recipients.

15. For details on I-Corps, NASA open innovation programs, and the University Innovation Fellows see Arkilic, "Raising the NSF Innovation Corps," Jenn Gustetic, "Innovation for Every American," and Humera Fasihuddin and Leticia Britos Cavagnaro, "An Innovator's Movement," in Wisnioski, Hintz, and Kleine, *Does America Need More Innovators?* 25–40, 69–82, 105–130.

16. White House, "Educate to Innovate," http://www.whitehouse.gov/issues/education/k-12/educate-innovate; A. L. Bement et al., *Educate to Innovate: Factors That Influence Innovation: Based on Input from Innovators and Stakeholders* (Washington, DC: The National Academies Press, 2015).

17. Barack Obama, "Remarks by the President in State of Union Address," The White House, January 25, 2011, https://obamawhitehouse.archives.gov/the-press-office/2011/01/25/remarks-president-state-union-address.

18. "Innovation: An American Imperative," American Academy of Arts and Sciences, June 23, 2015, https://cnsf.us/news/InnovationImperativeCallToAction.pdf.

19. Sebastian M. Pfotenhauer and Sheila Jasanoff, "Traveling Imaginaries: The 'Practice Turn' in Innovation Policy and the Global Circulation of Innovation Models," in *The Routledge Handbook of the Political Economy of Science*, ed. David Tyfield et al. (London: Routledge, 2017), 416–428.

20. Adam Karbowski et al., *Strengthening the Knowledge Base for Innovation in the European Union*, ed. Marzenna Weresa (Warsaw: Polish Scientific Publishers PWN, 2018); Sybille van den Hove et al., "The Innovation Union: A Perfect Means to Confused Ends?" *Environmental Science & Policy* 16 (2012): 73–80.

21. Lilly Irani, *Chasing Innovation: Making Entrepreneurial Citizens in Modern India* (Princeton, NJ: Princeton University Press, 2019); Silvia M. Lindtner, *Prototype Nation: China and the Contested Promise of Innovation* (Princeton, NJ: Princeton University Press, 2021).

22. Gabriela Dutrénit and Gustavo Crespi, *Science, Technology and Innovation Policies for Development: The Latin American Experience* (Berlin, Germany: Springer, 2014); Michael J. Kahn, "The Status of Science, Technology and Innovation in Africa," *Science, Technology and Society* 27, no. 3 (March 27, 2022): 327–350.

23. Richard Florida, *Rise of the Creative Class, Revisited*, 2nd ed. (New York: Basic Books, 2012), vii–xix; Charlynn A. Burd, "Metropolitan Migration Flows of the Creative Class by Occupation Using 3-Year 2006–2008 and 2009–2011 American Community Survey Data," Census.gov, April 2013, https://www.census.gov/library/working-papers/2013/demo/SEHSD-WP2013-11.html; Richard Florida, "Maps Reveal Where the Creative Class Is Growing," *Bloomberg*, July 9, 2019, https://www.bloomberg.com/news/articles/2019-07-09/maps-reveal-where-the-creative-class-is-growing.

24. Bruce Katz and Julie Wagner, *The Rise of Innovation Districts: A New Geography of Innovation in America* (Washington, DC: Metropolitan Policy Program at Brookings, 2014).

25. Eric von Hippel, *Democratizing Innovation* (Cambridge, MA: MIT Press, 2006); Henry W. Chesbrough, *Open Innovation: The New Imperative for Creating and Profiting from Technology* (Boston, MA: Harvard Business School Press, 2003).

26. Christensen, *The Innovator's Dilemma*.

27. Doreen Massey and David Wield, *High Tech Fantasies: Science Parks in Society, Science and Space* (Abingdon, UK: Routledge, 1992).

28. Pets.com, an online retailer of pet supplies, became a symbol for the internet economy's magical thinking. It spent millions of dollars on clever Super Bowl advertisements with a sock puppet mascot, despite annual revenue of less than a million dollars and selling all its products at a loss.

29. Philip Mirowski, *Science-Mart*; Richard Sennett, *The Corrosion of Character: The Personal Consequences of Work in the New Capitalism* (New York: Norton, 2011); Evgeny Morozov, *The Net Delusion: The Dark Side of Internet Freedom* (New York: PublicAffairs, 2012).

30. Evgeny Morozov, "Our Naive Innovation Fetish," *The New Republic*, March 17, 2014, https://newrepublic.com/article/116939/innovation-fetish-naive-buzzword-unites-parties-avoids-policy-choice.

31. Lee Vinsel, "95 Theses on Innovation," November 12, 2015, http://leevinsel.com/blog/2015/11/12/95-theses-on-innovation.

32. Thomas Frank, *Listen Liberal, or What Ever Happened to the Party of the People?* (New York: Metropolitan Books, 2016), 27–35, 70.

33. See e.g., Jacob Silverman, "The Crowdsourcing Scam: Why Do You Deceive Yourself?" *The Baffler*, October 2014, https://thebaffler.com/salvos/crowdsourcing-scam; Patrick Vitale, "The Pittsburgh Fairy Tale," *Jacobin*, June 20, 2017, https://jacobin.com/2017/06/pittsburgh-tech-new-economy-manufacturing-inequality.

34. Benoît Godin, Gérald Gaglio, and Sebastian Pfotenhauer, "X-Innovation: Re-Inventing Innovation Again and Again," *NOvation: Critical Studies of Innovation*, 1 (2019), 1–16.

35. The term itself dates to 2013 but became common after 2018. Adrian Wooldridge, "The Coming Tech-Lash," *The Economist* (November 18, 2013). https://www.economist.com/news/2013/11/18/the-coming-tech-lash. For examples of the genre, see Cathy O'Neil, *Weapons of Math Destruction: How Big Data Increases Inequality and Threatens Democracy* (New York: Crown, 2016); Jonathan Taplin, *Move Fast and Break Things: How Facebook, Google, and Amazon Cornered Culture and Undermined Democracy* (Boston, MA: Little, Brown, 2017); Franklin Foer, *World Without Mind: The Existential Threat of Big Tech* (London: Penguin Publishing Group, 2017); Safiya Umoja Noble, *Algorithms of Oppression: How Search Engines Enforce Racism* (New York: New York University Press, 2018); Jamie Bartlett, *The People vs Tech: How the Internet Is Killing Democracy (And How We Save It)* (London: Penguin, 2018); Ruha Benjamin, *Race after Technology: Abolitionist Tools for the New Jim Code* (Cambridge, UK: Polity, 2019).

36. Teresa Mull, "Big Tech Is the New Big Tobacco," *The Federalist*, August 31, 2022, https://thefederalist.com/2022/08/31/big-tech-is-the-new-big-tobacco/; J. B. Shurk, "Big Tech Slept with Communist China and Brought Venereal Censorship Back to America," *The Federalist*, October 22, 2020, https://thefederalist.com/2020/10/22/big-tech-slept-with-communist-china-and-brought-venereal-censorship-back-to-america/. Vinsel, "Design Thinking Is Kind of Like Syphilis."

37. Josh Hawley, *The Tyranny of Big Tech* (Washington, DC: Regnery Publishing, 2021).

38. Michelle Malkin and John Miano, *Sold Out: How High-Tech Billionaires & Bipartisan Beltway Crapweasels Are Screwing America's Best & Brightest Workers* (New York:

Mercury Ink, 2015); Jeremy Carl, "The Tech Amnesty," *National Review*, July 31, 2017, https://www.nationalreview.com/magazine/2017/07/31/silicon-valley-immigration-support-jobs/; Floyd Brown and Todd Cefaratti, *Big Tech Tyrants: How Silicon Valley's Stealth Practices Addict Teens, Silence Speech, and Steal Your Privacy* (New York: Simon & Schuster, 2019).

39. Donald J. Trump, "Presidential Memorandum on The White House Office of American Innovation," March 27, 2017, https://trumpwhitehouse.archives.gov /presidential-actions/presidential-memorandum-white-house-office-american -innovation/.

40. Aside from a 2018 report on modernizing federal information technology systems, the OIA did not register a single programmatic achievement. See Jessie Bur, "Innovation at Scale: What Has the White House Office of American Innovation Accomplished?" *Federal Times*, March 14, 2018, https://www.federaltimes.com/it-networks/2018/03 /14/innovation-at-scale-what-has-the-white-house-office-of-american-innovation -accomplished/.

41. Jeffrey Mervis, "Trump's White House Science Office Still Small and Waiting for Leadership," *Science*, July 11, 2017, doi: 10.1126/science.aan7084.

42. Nick Bilton, "Elon Musk's Totally Awful, Batshit-Crazy, Completely Bonkers, Most Excellent Year," *Vanity Fair*, November 10, 2020, https://www.vanityfair .com/news/2020/11/elon-musks-totally-awful-batshit-crazy-most-excellent-year; Sandra Song, "Watch Jeff Bezos' Laugh Go Full Supervillain over the Years," *PAPER Magazine*, July 28, 2021, https://www.papermag.com/jeff-bezos-laugh-tiktok; Liam Gaughan, "From Bond Villains to Mr. Burns, All the Famous Bad Guys Elon Musk Has Ripped Off," *Dallas Observer*, November 16, 2022, https://www.dallasobserver .com/arts/10-times-elon-musk-was-a-movie-villain-15277874.

43. See for example: Dan Lyons, *Disrupted: My Misadventure in the Start-up Bubble* (New York: Hachette books, 2016); García Antonio Martínez, *Chaos Monkeys: Obscene Fortune and Random Failure in Silicon Valley* (New York: Harper, 2016); Corey Pein, *Live Work Work Work Die: A Journey into the Savage Heart of Silicon Valley* (New York: Metropolitan Books, 2018); Anna Wiener, *Uncanny Valley: A Memoir* (New York: MCD Books, 2020).

44. Vinsel and Russell, *The Innovation Delusion*; Jonathan A. Knee, *The Platform Delusion: Who Wins and Who Loses in the Age of Tech Titans* (New York: Penguin Publishing Group, 2021); Eliot Brown and Maureen Farrell, *The Cult of We: WeWork and the Great Start-Up Delusion* (New York: Crown, 2021).

45. Frank, *Listen, Liberal*, 133.

46. Lilly Irani, "Design Thinking": Defending Silicon Valley at the Apex of Global Labor Hierarchies," *Catalyst: Feminism, Theory, Technoscience* 4, no. 1 (2018): 1–19. See also Natasha Iskander, "Design Thinking is Fundamentally Conservative and preserves the status quo," *Harvard Business Review*, September 5, 2018, https://hbr

.org/2018/09/design-thinking-is-fundamentally-conservative-and-preserves-the
-status-quo.

47. Jill Lepore, "The Disruption Machine: What the Gospel of Innovation Gets Wrong," *New Yorker*, June 23, 2014, http://www.newyorker.com/magazine /2014/06/23/the-disruption-machine.

48. See especially Winner, "The Cult of Innovation" and Vinsel, "95 Theses on Innovation."

49. Luis Suarez-Villa, *Technology and Oligopoly Capitalism* (New York: Routledge, 2023). For a broader discussion of rising inequality, see Thomas Piketty, *Capital in the Twenty-First Century*, trans. Arthur Goldhammer (Cambridge, MA: Harvard University Press, 2014).

50. Liza Mundy, "Why Is Silicon Valley so Awful to Women?" *The Atlantic*, April 2017, https://www.theatlantic.com/magazine/archive/2017/04/why-is-silicon-valley-so-awful-to-women/517788; Sheelah Kolhatkar, "The Disrupters," *The New Yorker*, November 13, 2017, https://www.newyorker.com/magazine/2017/11/20/the-tech-industrys-gender-discrimination-problem; Trae Vassallo et al., The Elephant in the Valley, accessed May 15, 2023, https://www.elephantinthevalley.com/.

51. Tyler Cowen, *The Great Stagnation: How America Ate All the Low-Hanging Fruit of Modern History, Got Sick, and Will (Eventually) Feel Better* (New York: Dutton, 2011).

52. Robert J. Gordon, *The Rise and Fall of American Growth: The U.S. Standard of Living Since the Civil War* (Princeton, NJ: Princeton University Press, 2016).

53. Ross Douthat, *Decadent Society: How We Became the Victims of Our Own Success* (New York: Avid Reader Press, 2020), 22, 36, 45–46.

54. Suzanne Berger, *Making in America: From Innovation to Market* (Cambridge, MA: MIT Press, 2013), 222. See also Vaclav Smil, *Made in the USA: The Rise and Retreat of American Manufacturing* (Cambridge, MA: MIT Press, 2013).

55. Christo Sims, *Disruptive Fixation: School Reform and the Pitfalls of Techno-Idealism* (Princeton, NJ: Princeton University Press, 2017).

56. David V. Johnson, "Innovator-In-Chief: Barack Obama, Venture Capitalist?" *The Baffler*, September 16, 2016, https://thebaffler.com/latest/innovator-in-chief-johnson.

57. Anand Giridharadas, *Winners Take All: The Elite Charade of Changing the World* (New York: Knopf, 2018). See also Raj Kumar, *The Business of Changing the World: How Billionaires, Tech Disrupters, and Social Entrepreneurs Are Transforming the Global Aid Industry* (Boston, MA: Beacon Press, 2019).

58. Matt Ridley, *How Innovation Works and Why It Flourishes in Freedom* (New York: HarperCollins, 2020).

59. The Brookings Institution followed up on its innovation district strategy with a study of the "dilemmas" and "dead zones" caused by their advocacy. Julie Wagner,

"Innovation Districts and Their Dilemmas with Place," Brookings, February 21, 2019, https://www.brokings.edu/blog/the-avenue/2019/02/21/innovation-districts -and-their-dilemmas-with-place/.

60. The idea of social innovation had existed in practice (if not in name) since the nineteenth-century utopian communities that inspired twentieth-century innovation experts. See chapters 2 and 7 in this book. See also Benoît Godin, "Social Innovation: Utopias of Innovation from c.1830 to the Present," Project on the Intellectual History of Innovation Working Paper No. 11, 2012, http://www.csiic.ca/PDF/SocialInnovation _2012.pdf; Noorseha Ayob, Simon Teasdale, and Kylie Fagan, "How Social Innovation 'Came to Be': Tracing the Evolution of a Contested Concept," *Journal of Social Policy* 45, no. 4 (2016): 635–653; Liliya Satalkina and Gerald Steiner, "Social Innovation: A Retrospective Perspective," *Minerva* 60 (2022): 567–591.

61. Ashoka Innovators for the Public, "What is a Public Entrepreneur?" brochure, box 2, Ashoka Innovators for the Public, Haas Center for Public Service Records, SC0541, Stanford University Special Collections and University Archives.

62. Gregory Dees and Alice Oberfield, *Note on Starting a Nonprofit Venture* (Boston, MA: Harvard Business School, 1991); Gregory J. Dees, "Social Enterprise: An Alternative Career Choice for MBAs," *Careers and the Minority MBA* (Fall 1997), 52–53; Erin L. Worsham, "Reflections and Insights on Teaching Social Entrepreneurship: An Interview with Greg Dees," *Academy of Management Learning & Education* 11, no. 3 (2012): 442–452.

63. The innovation think tank concluded that if left unchecked inequality would lead to domestic unrest that would diminish the nation's "growth and competitiveness." Fred O'Regan and Maureen Conway, *From the Bottom Up: Toward a Strategy for Income and Employment Generation among the Disadvantaged, an Interim Report* (Washington, DC: Aspen Institute, 1993).

64. Christine W. Letts, William P. Ryan, and Allen Grossman, *High Performance Nonprofit Organizations: Managing Upstream for Greater Impact* (New York: John Wiley & Sons, 1998), 169–192.

65. Paul Charles Light, *The Search for Social Entrepreneurship* (Washington, DC: Brookings Institution Press, 2008); M. Kim and J. Leu, "The Field of Social Entrepreneurship Education: From the Second Wave of Growth to a Third Wave of Innovation," in *Social Entrepreneurship Education Resource Handbook*, ed. Debbi D. Brock and Ashoka U (Washington, DC: Ashoka U, 2011), 4–5; Helen Coster, "Forbes' List of the Top 30 Social Entrepreneurs," *Forbes*, November 30, 2011, https://www.forbes .com/sites/helencoster/2011/11/30/forbes-list-of-the-top-30-social-entrepreneurs/; Rahim Kanani, "5 Brilliant Social Innovators You've Never Heard of, until Now," *Forbes*, February 17, 2014, http://www.forbes.com/sites/rahimkanani/2014/02/17/5 -brilliant-social-innovators-youve-never-heard-of-until-now/.

66. Michaela Jarvis, "Invention Ambassadors Take on Society's Challenges," *Science* 357, no. 6,353 (2017): 766–767.

67. Alex Bell, Raj Chetty, Xavier Jaravel, Neviana Petkova, and John Van Reenen, "Who Becomes an Inventor in America? The Importance of Exposure to Innovation," *The Quarterly Journal of Economics* 134, no. 2 (2019): 647–713. Economist Lisa Cook documented discrimination at each stage of the innovation economy from education to employment and patenting. See Lisa Cook, "Innovation Gap in Pink and Black," in Wisnioski, Hintz, and Kleine, *Does America Need More Innovators?* 221–247.

68. Debbie Chachra, "Why I Am Not a Maker," *The Atlantic*, January 23, 2015, https://www.theatlantic.com/technology/archive/2015/01/why-i-am-not-a-maker /384767/; "The Maker Movement Gets a Dose of Critique," *Remake Learning*, 2015, https://remakelearning.org/stories/the-maker-movement-gets-a-dose-of-critique/.

69. "An Open Letter from Women of Science," 500 Women Scientists, November 17, 2016, https://500womenscientists.org/our-pledge.

70. Londa Schiebinger, ed., *Gendered Innovations in Science and Engineering* (Palo Alto, CA: Stanford University Press, 2008); Gendered Innovations, http://genderedinnovations.stanford.edu.

71. Responsible innovation developed in parallel and in exchange with American practitioners in Europe, where it achieved widespread implementation. In 2010, the European Union invested tens of millions of dollars in government mandates for responsible innovation methods in R&D decision-making. René von Schomberg and Jonathan Hankins, eds. *The International Handbook of Responsible Innovation: A Global Resource* (Chatham, MA: Edward Elgar, 2019).

72. Richard Owen et al., "A Framework for Responsible Innovation," in *Responsible Innovation: Managing the Responsible Emergence of Science and Innovation in Society*, ed. Richard Owen and John Bessant (New York: Wiley, J, 2013), 27–50. This emphasis on responsibility, of course, was not entirely novel. Rogers's *Diffusion of Innovations* is again an instructive touch point. In the book's 1971 edition, he had added a chapter on consequences and the change agent's responsibility for them. By the fifth edition in 2003, the analysis of consequences had grown to include case studies and instructions for would-be innovators to evaluate the equitable outcomes of an innovation before undertaking it. See Everett M. Rogers, *Diffusion of Innovations*, 5th ed. (New York: Free Press, 2003).

73. David H. Guston, "Understanding 'Anticipatory Governance,'" *Social Studies of Science* 44, no. 2 (2013): 218–242; Vincent Blok, "Look Who's Talking: Responsible Innovation, the Paradox of Dialogue and the Voice of the Other in Communication and Negotiation Processes," *The Simple Care of a Hopeful Heart* 1, no. 2 (2014): 171–190.

74. Responsible innovation built on decades of efforts to proactively address technologies unwanted effects. In the 1960s and 1970s, this "technology assessment" emerged out of the field of science and technology studies and was institutionalized

in the Office of Technology Assessment. While Congress later eliminated the Office of Technology Assessment, high-profile initiatives on the human genome and nanotechnology renewed efforts to predict and manage the "societal implications" of technology. David H. Guston, Megan Jones, and Lewis M. Branscomb, "Technology Assessment in the U.S. State Legislatures," *Technological Forecasting and Social Change* 54, no. 2–3 (1997): 233–250.

75. W. Patrick McCray, "Will Small Still Be Beautiful: Making Policies for Our Nanotech Future," 21, no. 2 *History and Technology* (2005): 177–203.

76. W. Patrick McCray, *The Visioneers: How a Group of Elite Scientists Pursued Space Colonies, Nanotechnologies, and a Limitless Future* (Princeton, NJ: Princeton University Press, 2012).

77. Daniel Sarewitz, Advancing the Science of Science and Innovation Policy, *Testimony before the US House of Representatives Committee on Science and Technology; CSPO Report* (2010): 10–04, http://docs.politicascti.net/documents/Teoricos/Sarewitz_et_al.pdf.

78. "Align Technology with Humanity's Best Interests," Center for Humane Technology, accessed May 16, 2023, https://www.humanetech.com/.

79. Station1, https://www.station1.org.

80. Sasha Costanza-Chock, *Design Justice: Community-Led Practices to Build the Worlds We Need* (Cambridge, MA: MIT Press, 2020), 69–102.

81. See chapter 6 in this book.

82. Matt Ratto and Megan Boler, eds., *DIY Citizenship: Critical Making and Social Media* (Cambridge, MA: MIT Press, 2014); Arturo Escobar, *Designs for the Pluriverse: Radical Interdependence, Autonomy, and the Making of Worlds* (Durham, NC: Duke University Press, 2018); Ashley Shew, *Against Technoableism: Rethinking Who Needs Improvement* (New York: W.W. Norton, 2023); Carl DiSalvo, *Adversarial Design* (Cambridge, MA: MIT Press, 2012).

83. The Allied Media Conference began in 1999 as a meeting for underground Zine culture. It relocated to Detroit in 2007.

84. Joi Ito and Ethan Zackerman, "Announcing the Winners of the 2017 Media Lab Disobedience Award," MIT Media Lab, July 20, 2017, https://www.media.mit.edu/posts/reflections-on-the-disobedience-award/. The Prize was only offered twice and was then canceled when financier Jeffery Epstein's support of the Media Lab became a scandal.

85. Costanza-Chock had been an AMC member and advisor since the 2000s, when they were a graduate student, and became a board member in 2012. At MIT, they established the Codesign Toolkit. *Streets.* They were one of the judges of the Disobedience Prize.

86. Allied Media. "Book Sasha Costanza-Chock." https://alliedmedia.org/post/book-sasha-costanza-chock.

87. Design Justice Network, "Design Justice Network Principles," https://designjustice .org/read-the-principles.

88. Steven J. Jackson, "Rethinking Repair," in *Media Technologies: Essays on Communication, Materiality, and Society*, ed. Tarleton Gillespie, Pablo J. Boczkowski, and Kirsten A. Foot (Cambridge, MA: The MIT Press, 2014), 221–240; David L. Edgerton, *The Shock of The Old: Technology and Global History since 1900* (London: Profile Books, 2011).

89. Andrew Russell and Lee Vinsel, "Hail the Maintainers" *Aeon* (2016), https://aeon .co/essays/innovation-is-overvalued-maintenance-often-matters-more.

90. Costanza-Chock, *Design Justice*, 218–219.

91. See, for example, Audley Genus and Marfuga Iskandarova, "Responsible Innovation: Its Institutionalisation and a Critique," *Technological Forecasting and Social Change* 128 (2018): 1–9.

92. Vinsel and Russell, *The Innovation Delusion*, 50, 141–142, 144, 157–158. The book's publisher, Currency, an imprint of Doubleday, was a leading source of innovation toolkits, including Tom Kelley's and Jonathan Litman's *The Ten Faces of Innovation: IDEO's Strategies for Beating the Devil's Advocate and Driving Creativity Throughout Your Organization*, and Peter Thiel's and Blake Masters's *Zero to One*.

93. Navi Radjou, Jaideep C. Prabhu, and Paul Polman, *Frugal Innovation: How to Do Better with Less* (London: The Economist, 2015); "Make Business a Force for Good," B Lab, https://www.bcorporation.net/en-us/.

94. Mary Gray, "Just Tech: Centering Community-Driven Innovation at the Margins Episode 3 with Dr. Sasha Costanza-Chock," Microsoft Research, April 13, 2022, https://www.microsoft.com/en-us/research/podcast/just-tech-centering-community -driven-innovation-at-the-margins-episode-3-with-dr-sasha-costanza-chock/.

95. Ro Khanna, *Dignity in a Digital Age: Making Tech Work for All of Us* (New York: Simon & Schuster, 2022). See also Rob Reich, Mehran Sahami, and Jeremy Weinstein, *System Error: Where Big Tech Went Wrong and How We Can Reboot* (London: Hodder & Stoughton, 2021).

96. Biden tapped science and technology studies scholar Alondra Nelson to lead OSTP and appointed economist Lisa Cook to the Federal Reserve Board of Governors.

97. "Remarks By President Biden on the Bipartisan Innovation Act," The White House, May 6, 2022, https://www.whitehouse.gov/briefing-room/speeches-remarks /2022/05/06/remarks-by-president-biden-on-the-bipartisan-innovation-act/; Jeffery Mervis, "U.S. Innovation Bill Clears Major Senate Hurdle with Research Provisions Intact," *Science*, July 21, 2022, doi: 10.1126/science.ade0590; Mark Muro, "Can the Chips Act Heal the Nation's Economic Divides?" Brookings, August 2, 2022, https:// www.brookings.edu/blog/the-avenue/2022/08/02/can-the-chips-act-heal-the-nations -economic-divides/.

98. National Science Foundation, "NSF Invests more than $43 million in NSF Regional Innovation Engines Development Awards," May 11, 2023, https://new.nsf .gov/funding/initiatives/regional-innovation-engines/updates/nsf-invests-more-43 -million-nsf-regional.

99. "PUBLIC LAW 117–58," Congress.gov, November 15, 2021, https://www.congress .gov/117/plaws/publ58/PLAW-117publ58.pdf.

100. "Biden-Harris Administration Announces Historic Actions to Advance National Vision for STEMM Equity and Excellence," The White House, December 12, 2022, https://www.whitehouse.gov/ostp/news-updates/2022/12/12/biden-harris -administration-announces-historic-actions-to-advance-national-vision-for-stemm -equity-and-excellence/.

101. Rebecca Nelson and *National Journal*, "The Secret Republicans of Silicon Valley," *The Atlantic*, April 8, 2015, https://www.theatlantic.com/politics/archive/2015/04 /the-secret-republicans-of-silicon-valley/451086; Karen Weise, "The Lonely Lives of Silicon Valley Conservatives," *Bloomberg*, September 5, 2017, https://www.bloomberg .com/news/articles/2017-09-05/the-lonely-lives-of-silicon-valley-conservatives.

102. Laurie Segall, "The New Counterculture: Conservatives in Silicon Valley?" *CNNMoney*, November 10, 2017, https://money.cnn.com/2017/11/10/technology /culture/divided-we-code-undercover-conservatives/index.html.

103. Joel N. Shurkin, *Broken Genius: The Rise and Fall of William Shockley, Creator of the Electronic Age* (London: Palgrave Macmillan, 2006).

104. Max Chafkin, *The Contrarian: Peter Thiel and Silicon Valley's Pursuit of Power* (London: Bloomsbury Publishing, 2021), 40, 75–76.

105. Bruce Gibney, "What Happened to the Future?" Founders Fund, https:// foundersfund.com/the-future/.

106. Peter Thiel and Blake Masters, *Zero to One: Notes on Startups, or How to Build the Future* (New York: Currency, 2014), 173–189.

107. Morgan Brinlee, "Peter Thiel's Speech Transcript Has a Big Moment," *Bustle*, July 21, 2016, https://www.bustle.com/articles/174180-the-transcript-of-peter-thiels -speech-includes-a-historic-lgbtq-rights-moment.

108. Chafkin, *The Contrarian*, 251–254.

109. Benjamin Wallace-Wells, "The Rise of the Thielists," *The New Yorker*, May 13, 2021, https://www.newyorker.com/news/annals-of-populism/the-rise-of-the -thielists.

110. Vance, *Elon Musk*.

111. Alyssa Bereznak and Kate Knibbs, "The Great Elon Musk Debate," *The Ringer*, September 13, 2017, https://www.theringer.com/tech/2017/9/13/16302426/elon -musk-villain-hero; Elizabeth Lopatto, "Why So Many People Think Elon Musk Is a

Hero—or a Villain," *The Verge*, November 30, 2018, https://www.theverge.com/2018/11/30/18118414/elon-musk-hero-villain-genius-spacex-tesla-boring-company.

112. Joe Rogan, interview with Elon Musk, *Joe Rogan Experience*, podcast audio, September 7, 2018.

113. Molly Ball, Jeffrey Kluger, and Alejandro de la Garza, "Elon Musk: *Time* 2021 Person of the Year," *Time*, December 13, 2021, https://time.com/6127757/elon-musk-interview-person-of-the-year-2021/.

114. Rich Lowry, "In Defense of Elon Musk," *National Review*, April 19, 2022, https://www.nationalreview.com/2022/04/in-defense-of-elon-musk-2/.

115. Donna Edmunds, "Elon Musk's Vision for the Future Is an Inspiration for Dreamers Everywhere," Heroes of Liberty, December 2, 2022, https://heroesofliberty.com/blogs/heroes-of-literacy/elon-musk-s-vision-for-the-future-is-an-inspiration-for-dreamers-everywhere.

116. Vivek Ramaswamy, *Woke, Inc: Inside Corporate America's Social Justice Scam* (New York: Center Street, 2021).

117. Vivek Ramaswamy, "America First 2.0," https://www.vivek2024.com/america-first-2-0/.

118. Thiel Fellowship, https://thielfellowship.org. Max Chafkin argues that the selected fellows were "nearly all of them—boys, and, almost to a person, they shared Thiel's social awkwardness." Chafkin, *The Contrarian*, 161–169. At the same time, one of the initiative's successful alumna is Madison Maxey, a Black woman entrepreneur whose start-up produces electronic textiles, who was also profiled by the Smithsonian's Lemelson Center for its *Innovative Lives* series and profiled in Heather Cabot and Samantha Walravens, *Geek Girl Rising: Inside the Sisterhood Shaking Up Tech* (New York: St. Martins, 2017).

119. Synthesis, https://www.synthesis.com/.

120. Ayana Elizabeth Johnson, "How to Find Joy in Climate Action," TED Talk, 2022, https://www.ted.com/talks/ayana_elizabeth_johnson_how_to_find_joy_in_climate_action/transcript.

121. "Circles," The All We Can Save Project, https://www.allwecansave.earth/circles.

CHAPTER 10

1. *Innovation* is following a similar pattern to the formerly exalted concepts of *efficiency* and *progress*. Like innovation, both experienced intense support and sharp falls from grace, but both continue to shape human experience. *Efficiency* is the more modest comparison. Like innovation, it was the product of novel experts. At the turn of the twentieth century, these experts claimed to have discovered solutions that addressed both the technological and societal challenges of a modernizing world.

Frederick Taylor, efficiency's most recognizable champion, called his approach "scientific management" and promised that techniques for the factory floor applied to good government. Meanwhile, democratic and communist regimes embraced efficiency with the hopes of making a profit and improving lives. As negative consequences mounted, however, critics attacked efficiency as a soulless and delusional corporate tool and called its acolytes a cult. But the ideas and practices of efficiency were built into industrial society's material and psychological infrastructure. Reformers worked for decades to develop humanistic variations, and efficiency remains omnipresent in global supply chains and individual "life-hacks." Its benefits are self-evident, as are its downsides. Innovation's relationship to *progress* is more complex. While ancient in origin, *progress* rose in the eighteenth century as a watchword of the "scientific revolution," the "industrial revolution," and "Enlightenment." Despite attracting enemies from the start, progress has served as a structuring idea for centuries, as a rationalization for colonial expansion and environmental destruction, but also liberation and justice. In World War II's aftermath, progress came under fierce critique. Its rosy claims about the inherent benefits of science and technology became especially circumspect. Intellectuals declared the "end of enlightenment" and the "death of nature." Innovation's ascension was a direct outcome of progress's decline. Just as there is no "after efficiency" or "after progress," there will be no after innovation.

2. Gardner, *Self-Renewal*, xi.

INDEX